"An ominous foreign presence suddenly seems to take control of the skies—'Another Pearl Harbor!' some shout. Initial fears are replaced by a determination to meet the challenge, and America declares that life has been changed forever. Sounds familiar, but the transforming event of Paul Dickson's book is not the crash of hijacked airliners [on] Sept. 11; it is the Soviet Union's launch in October 1957 of Sputnik."
—*The Washington Post Book World*

"Delightful . . . Relying on newly declassified Cold War documents and his storytelling skill, Dickson entertainingly illuminates the post-Sputnik turmoil that led not just to flights to the moon and planets but also to altering vast swaths of American life, from education to computers to civil rights and feminism. He captures perfectly the frantic, even zany, flavor of the times and the dizzying social impact of the 'Red Moon.' Even the Internet, Dickson says, owes its genesis to the panic over Sputnik. . . . Readers will encounter surprises. . . . An engaging account of the Sputnik revolution."
—*The Sunday Oregonian*

"The best popular book about Sputnik."
—*Kansas City Star*

"*Sputnik* should climb far up the lists, and have a long ride." —*The Baltimore Sun*

"At 184 pounds, Sputnik was scarcely larger than a basketball and did nothing except emit beeps out of a battery-powered antenna. That beep brought about the only direct hardship America suffered from Sputnik: It reportedly activated electric garage-door openers from coast to coast. . . . [*Sputnik: The Shock of the Century* is] a straightforward and snappy account of a crisis in American politics, science, and self-esteem."
—*New York Observer*

"This account of the beginnings of the space race should be required reading for those who were born after 1960. It will be a joy to read for those who remember the beep-beep of a small metal ball circling the earth—who remember standing in the cold late autumn night hoping to see Sputnik as it passed over our part of Earth."
—*The Roanoke Times & World News*

"[A] solid overview."
—*The Cleveland Plain Dealer*

"Valuable contributions to the historical record."
—*The National Journal*

"Breezily written."
—*USA Today*

continued . . .

"Dickson has written a terrific book about Sputnik."

—*Bulletin of the Atomic Scientists*

"Dickson vividly describes the events leading up to Sputnik and its immediate and long-term impact of the United States." —*The Charlotte Observer*

"Captures the excitement and angst of the dawning of the Space Age."

—*The Dallas Morning News*

"Absorbing . . . the best book on the political shockwaves following Sputnik."

—*New Scientist*

"Compelling reading . . . *Sputnik* has a handful of great strengths."

—*Toronto Globe and Mail*

"*Sputnik* is a surprise. While the subject matter suggests it is basically for the scientifically minded, the book has an important message for the layman. . . . Dickson's account of all the events in the space race is both accurate and exciting." —*Richmond Times-Dispatch*

"Any history buff interested in knowing the scientific background of our present communication system will find the book fascinating. So will those who wonder about our American reaction in a threatening situation such as the Cold War and the Space Age and, perhaps, what paths we should take in our present world situation." —*The Sunday Oklahoman*

"A book suddenly made timely by the shock of this new century."

—*The Seattle Post-Intelligencer*

"[Dickson] completely understands the lure and lore of Sputnik."

—*Publishers Weekly*

"Fascinating . . . Dickson makes the space race come alive in layman's language, and he shows how the shock of the Russians being first at something galvanized this country in all sorts of far-reaching ways." —*Booklist*

"Sputnik is a fascinating, fast-paced, well-researched book. For post–September 11 Americans who now search the skies at every roar of a low-flying aircraft, these events of forty-four years ago resonate with urgency and, strangely enough, seem prophetic. Like the best social and scientific histories, Dickson's look back in time sheds a clearer beam on the road ahead." —Barbara A. Genco,

School Library Journal, from her list of the 10 best "grown up" books for 2001

"An excellent treatment of one of the early chapters of the Cold War."

—*Kirkus Reviews*

"Dickson re-creates the fire, furor, frustration, and flamboyance of the early Space Age. Sputnik's arrival set off a tidal wave in the affairs of men."

—Vinton Cerf, co-inventor of the Internet

"Dickson's book not only presents a thoughtful analysis of the impact Sputnik had on the dawning of the Space Age, but also serves as a valuable resource for understanding the historical context of the debates now taking place on issues such as National Missile Defense and the future of space."

—Susan Eisenhower, president of the Eisenhower Institute

"An insightful look at the way Sputnik changed the world, especially the United States—boosting its education and research. Sputnik essentially awakened America and started its space era." —Sergei N. Khrushchev,

author of *Nikita Khrushchev: Creation of a Superpower*

and senior fellow of the Watson Institute of International Studies

"Well written and informative, the book is a magnificent assessment of Cold War history . . . Dickson puts Sputnik into clear perspective—its impact on the course of international diplomacy, the growth of technology, and the hopes and fears of ordinary people in the U.S. and throughout the world."

—Francis Gary Powers, Jr.,

founder of the Cold War Museum

"Entertaining, admirably straightforward." —*Wilson Quarterly*

"Absolutely compelling." —*Metrowest Daily News*

"Fascinating . . . a sharply focused snapshot of a nation caught with its jaw dropped, partly in fear and partly in wonder at the marvel of a new era."

—*The Philadelphia Inquirer*

SELECTED BOOKS BY PAUL DICKSON

Think Tanks (1971)
The Great American Ice Cream Book (1972)
The Electronic Battlefield (1976)
The Future of the Workplace (1976)
The Future File (1977)
Out of This World: American Space Photography (1977)
The Official Rules (1978)
On Our Own: A Declaration of Independence for the Self-Employed (1985)
The Library in America (1986)
Family Words (1988)
The New Official Rules (1989)
Slang (1990)
Timelines (1990)
Baseball's Greatest Quotations (1991)
On This Spot: Pinpointing the Past in Washington, D.C. (with Douglas E. Evelyn) (1992)
Baseball: The President's Game (with William B. Mead) (1993)
The Congress Dictionary (with Paul Clancy) (1993)
Myth-Informed (with Joseph C. Goulden) (1993)
The Joy of Keeping Score (1996)
What's in a Name (1996)
The New Dickson Baseball Dictionary (1999)

Sputnik

THE SHOCK OF THE CENTURY

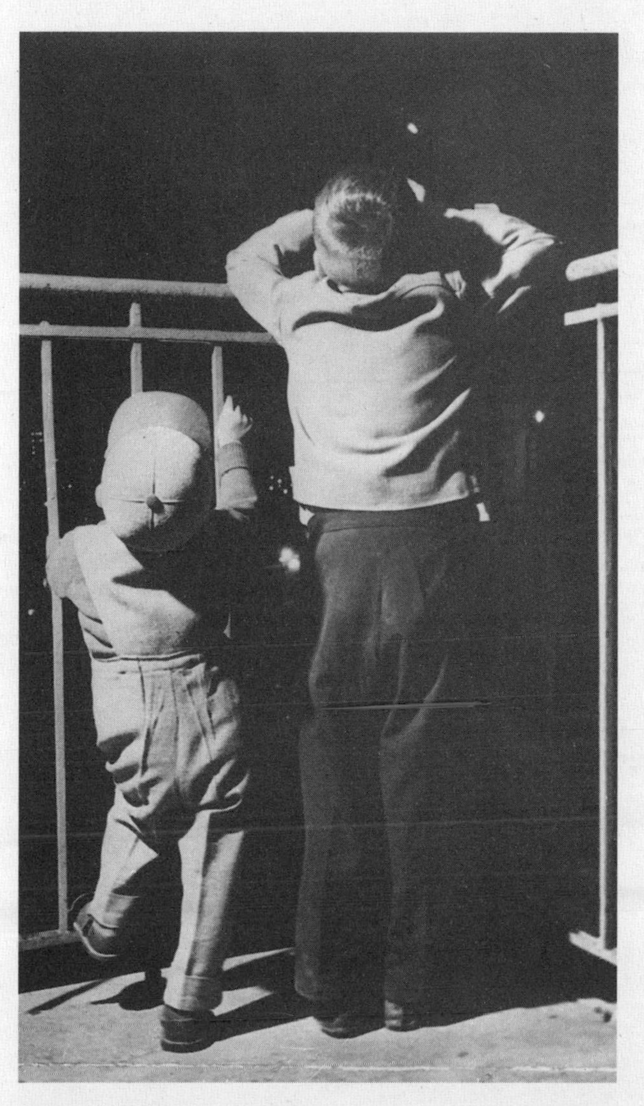

BERKLEY
BOOKS
NEW YORK

Paul Dickson

A Berkley Book
Published by The Berkley Publishing Group
A division of Penguin Putnam Inc.
375 Hudson Street
New York, New York 10014

Published by arrangement with Walker Publishing Company, Inc.

Cover and title page image: This image is part of a set of thirty-seven captioned photographs received by a North Dakota schoolteacher in 1959 after he wrote to the Soviet Union for information on the early days of the Sputniks. Its caption read "They want to be the first to spot Sputnik III." The collection was posted on eBay in November 1998, and purchased by the author. Other photos from this collection that appear throughout the book are credited as author's collection.
Cover design by Jill Boltin
Cover photo by NASA

PRINTING HISTORY
Walker & Company hardcover edition / October 2001
Berkley trade paperback edition / April 2003

Library of Congress Cataloging-in-Publication Data

Dickson, Paul.
Sputnik : the shock of the century / Paul Dickson.
p. cm.
Originally published: New York : Walker Pub., 2001.
Includes bibliographical references and index.
ISBN 0-425-18843-4
1. Sputnik satellites—History. 2. Artificial satellites, Russian—Political aspects. 3. Astronautics and state—United States—Public opinion. 4. United States.—Politics and government—20th century. I. Title.

TL796.5.S652 S664 2003
629.46'0947—dc21 2002027759

Printed in the United States of America

10 9 8 7 6 5 4 3 2 1

From the vantage point of 2100 A.D., the year of 1957 will most certainly stand in history as the year of man's progression from a two-dimensional to a three-dimensional geography. It may well stand, also, as the point in time at which intellectual achievement forged ahead of weapons and national wealth as instruments of national policy. The earth satellite is a magnificent expression of man's intellectual growth—of his ability to manipulate to his own purposes the very laws that govern his universe.

—Physicist Lloyd V. Berkner, October 1957

Contents

Introduction

> Never before had so small and so harmless an object created such consternation.
>
> —Daniel J. Boorstin, *The Americans: The Democratic Experience*

"Listen now," said the NBC radio network announcer on the night of October 4, 1957, "for the sound that forevermore separates the old from the new." Next came the chirping in the key of A-flat from outer space that the Associated Press called the "deep beep-beep." Emanating from a simple transmitter aboard the Soviet Sputnik satellite, the chirp lasted three-tenths of a second, followed by a three-tenths-of-a-second pause. This was repeated over and over again until it passed out of hearing range of the United States.

The satellite was silver in color, about the size of a beach ball, and weighed a mere 184 pounds. Yet for all its simplicity, small size, and inability to do more than orbit the Earth and transmit meaningless radio blips, the impact of Sputnik on the United States and the world was enormous and unprecedented.

The vast majority of people living today, at the beginning of the twenty-first century, were born after Sputnik was launched and may be unaware of the degree to which it helped shape life as we know it. Now is an especially good time to take a fresh and focused look at the event whose impact looms even larger with the passing of time. In the last decade an incredible amount of once-secret material has been declassified and made public. Scholars and writers both inside and outside government have coaxed key

Cold War documents out of hiding. Collectively, this material has given new dimensions and twists to almost every aspect of the events leading up to and following the launch of Sputnik.

For example, one recently released document reveals evidence of a long-forgotten pre-Sputnik "olive branch" extended by Russian scientists, who asked their American counterparts to supply a piece of scientific equipment for a planned launch. By most indications, this piece of equipment was meant for the third Sputnik.

It is not widely known even now that one of the reasons President Dwight D. Eisenhower and those around him did not react with alarm over Sputnik going into space ahead of an American satellite was that Eisenhower welcomed the launch to help establish the principle of "freedom of space." At the time of the Sputnik "crisis," the White House, Central Intelligence Agency, Air Force, and a few highly select and trustworthy defense contractors were creating a spy satellite that was so secret that only a few dozen people knew of it. Even its name, CORONA, was deemed secret for many years. Instead of being concerned with winning the first round of the space race, Eisenhower and his National Security Council were much more interested in launching surveillance satellites that could tell American intelligence where every Soviet missile was located.

For many of us born before the 1950s, the fascination and astonishment engendered by the launch of Sputnik remain fresh in our minds. Like many of my generation, I can recall exactly where I was when I heard about Sputnik's launch. I was eighteen years old, a college freshman at Wesleyan University in Middletown, Connecticut. A friend stopped me in the middle of the campus to say that he had heard about it on the radio. Instinctively, we both looked up.

Within hours I would actually hear its signal rebroadcast on network radio. Before the weekend was over, I got to hear it directly on a shortwave radio as it passed overhead.

Not only could you hear Sputnik, but, depending on where you were, it was possible to see it with the naked eye on certain days in the early morning or the late evening when the Sun was still close enough to the horizon to illuminate it.* While standing in the middle of the college football field a week or so after the launch, I first saw the satellite scooting across a dark

*What most of us who saw Sputnik in orbit would not learn for decades was that we probably were looking at the orbiting booster and not the satellite itself.

evening sky orbiting the Earth at a speed of 18,000 miles per hour. Watching Sputnik traverse the sky was seeing history happen with my own eyes. To me, it was as if Sputnik was the starter's pistol in an exciting new race. I was electrified, delirious, as I witnessed the beginning of the Space Age.

Prior to Sputnik, popular interest in science and technology had been on the rise since as early as the 1939 "World of Tomorrow" World's Fair in Flushing Meadows, New York. I attended the fair, albeit in utero, as I was born three days after my parents' last visit. But they saved many artifacts of the fair for me, including an official guidebook, which fascinated me as a kid and jump-started my interest in all sorts of things, particularly space travel.

That guidebook turned out to be a preview of the future. Exhibits like Ford's "Road of Tomorrow," General Motors' "Futurama," and the multi-sponsored "Town of Tomorrow" were more than fanciful prototypes; many of their imagined advances made their way into everyday life within a couple of decades. The fair's centerpiece was "Democracity," and it heralded superhighways, ranch-style houses, rec rooms, workshops for "do-it-yourselfers," and booming suburbs (known as "satellites" in the Democracity display) replete with prefab houses, two-car garages, and stereophonic sound. Something called "television" was actually demonstrated at the RCA exhibit.

The Transportation Pavilion, devoted to space exploration, showed a rocketport, a moonport, and a rocketship shot from a "rocketgun." In one lavish demonstration you could simulate blastoff on a trip to Venus, then stroll a primeval jungle inhabited by Venusian beasts and a colony of Martians. The fair promised a day when sleek vehicles would take passengers to the planets as easily as they could fly from New York to Chicago. It was as if this orderly march into the future was a part of America's destiny.

As it turned out, the real "world of tomorrow" was delayed because of World War II, but its vision was carried intact into the late 1940s and early 1950s, when it began to be realized. Americans who had struggled through the Great Depression and the war embraced the promise of a burgeoning middle class having goods, services, and comforts that formerly had been the province of European royalty. The average family's car had more pure horsepower than existed in all the stables of Buckingham Palace a generation earlier.

By 1957, a new world was at hand for the United States. The country was creating an interstate highway system; the suburbs were growing; families with two cars and color televisions were becoming the norm. The highest peacetime federal budget in history ($71.8 billion) was in place, and it was the first year in which more than one thousand computers would be built, bought, and shipped. There were advances in public health, although none more stunning than Dr. Jonas Salk's discovery of a vaccine against polio, the scourge of an entire generation of children.

At the same time, social changes were beginning to transform the United States. A great struggle to achieve a more egalitarian society was beginning. The first civil rights legislation since Reconstruction had been enacted in Congress on September 9, less than a month before Sputnik's launch. The Arkansas National Guard was in Little Rock, Arkansas, enforcing the right of blacks to go to school with whites. Culturally, as well, the country was moving to a different beat. Rock 'n' roll had come onto the scene, and Elvis Presley owned the summer of 1957 with his two-sided monster hit record of "Don't Be Cruel" and "Hound Dog."

Just when Americans were feeling self-confident and optimistic about the future, along came the crude, kerosene-powered Sputnik launch. The space race was under way, and the Soviets had won the first leg—the United States was agog and unnerved.

"No event since Pearl Harbor set off such repercussions in public life," wrote historian Walter A. McDougall in *The Heavens and the Earth—A Political History of the Space Age.* Simon Ramo, space pioneer and cofounder of Thompson Ramo Woolridge, later known as TRW, Inc., wrote in *The Business of Science* that "the American response to the accomplishment of the Soviet Union was comparable to the reaction I could remember to Lindbergh's landing in France, the Japanese bombing of Pearl Harbor and Franklin D. Roosevelt's death."

There was a sudden crisis of confidence in American technology, values, politics, and the military. Science, technology, and engineering were totally reworked and massively funded in the shadow of Sputnik. The Russian satellite essentially forced the United States to place a new national priority on research science, which led to the development of microelectronics—the technology used in today's laptop, personal, and handheld computers. Many essential technologies of modern life, including the Internet, owe their early development to the accelerated pace of applied research triggered by Sputnik.

On another level, Sputnik affected national attitudes toward conspicuous consumption, as well, symbolically killing off the market for the Edsel automobile and the decadent automotive tail fin. It was argued that the engineering talents of the nation were being wasted on frivolities. Americans, wrote historian Samuel Flagg Bemis from the vantage point of 1962, "had been experiencing the world crisis from soft seats of comfort, debauched by [the] mass media . . . , pandering for selfish profit to the lowest level of our easy appetites, fed full of toys and gewgaws, our power, our manpower softened in will and body in a climate of amusement."

Sputnik also changed people's lives in ways that filtered into modern popular culture. Sputnik was the instrument that gave Stephen King the "dread" that fuels his novels, caused the prolific Isaac Asimov to begin calling himself a science writer rather than a science fiction writer, inspired Ross Perot to create an electronics dynasty, and led others to become cosmonauts and astronauts.

NASA astronaut Franklin R. Chang-Dìaz is a case in point. He was born on April 5, 1950, in San José, Costa Riça. On a trip to Venezuela in October 1957, the seven-year-old was told by his mother to look skyward to see the Russian satellite crossing the night sky. Although the young Franklin could not spot Sputnik, he became so infatuated with the fact that human influence had moved into space that he decided then and there that this was his future. Once the American manned space program was under way, he wrote to Wernher von Braun, director of the George C. Marshall Space Flight Center, to find out how he might apply to become an astronaut. In the form letter that came back, he was advised to get a scientific or engineering degree and learn to fly. He also was told that he would have to become an American citizen. The United States, after all, was in a race with the Soviet Union. At eighteen he came to the United States from Costa Rica; he received a bachelor of science degree in mechanical engineering from the University of Connecticut in 1973 and a doctorate in physics from the Massachusetts Institute of Technology in 1977. Along the way he became a U.S. citizen and then in 1981 an astronaut.* Chang-Dìaz hopes to go to Mars eventually.

Politically, Sputnik created a perception of American weakness, complacency, and a "missile gap," which led to bitter accusations, resignations of key military figures, and contributed to the election of John F. Kennedy, who emphasized the space gap and the role of the Eisenhower-Nixon adminis-

*As of this writing, he is a veteran of six space flights (STS 61-C in 1986, STS-34 in 1989, STS-46 in 1992, STS-60 in 1994, STS-75 in 1996, and STS-91 in 1998) and has logged over 1,269 hours in space. STS-91 Discovery (June 2–12, 1998) was the ninth and final Shuttle-Mir docking mission and marked the conclusion of the highly successful joint U.S.-Russian program.

tration in creating it. But although the Sputnik episode publicly depicted Eisenhower as passive and unconcerned, he was fiercely dedicated to averting nuclear war at a time when the threat was very real. His concern for national security took precedence over any concerns about beating the Russians into Earth orbit.

When Kennedy as president decided to put Americans on the Moon, he did so with the belief that voters who had been kids at the time of Sputnik were more willing than their parents to pay the high price of going into space.

Diplomatically, Sputnik helped realign the United States and Great Britain as allies. For a decade, ties between the two nations had weakened partly due to the 1946 Atomic Energy Act, which had deprived the United Kingdom of American nuclear secrets, and partly because of the strong position that the United States had taken against the British and French during the Suez Crisis, which had been prompted by Egypt's seizure of the Suez Canal in July 1956. Now with the common threat of Soviet power implied by Sputnik, NATO was strengthened, guaranteeing the placement of American nuclear arms in Europe. The satellite touched off a superpower competition that may well have acted as a surrogate contest for universal power—perhaps even a stand-in for nuclear world war.

NASA chief historian Roger D. Launius wrote on the fortieth anniversary of the launch: "To a remarkable degree, the Soviet announcement changed the course of the Cold War. . . . Two generations after the event, words do not easily convey the American reaction to the Soviet Satellite." Without Sputnik, it is all but certain that there would not have not been a race to the Moon, which became the centerpiece contest of the Cold War.

From the outset, wrangling among the branches of the military over control of the rockets that would take the United States into space threatened the success of the American space program even before Sputnik. Eisenhower was at odds with his generals over the program, and each branch of the service had its own aspirations of going into space. The main event pitted the Army's Wernher von Braun and his Rocket Team in Huntsville, Alabama, against a team from the Naval Research Laboratory. The Army had the mighty Jupiter C rocket and its own Orbiter or Deal satellite (later to become Explorer) pitted against the Navy's experimental Viking rocket and Vanguard satellite.

The most powerful early rockets were developed as weapons—first as German V-2 technology from World War II and ultimately as intercontinental ballistic missiles. The space program seemed destined for civilian control just as the power of the atomic bomb had been taken from the military a decade earlier. The National Aeronautics and Space Administration began in 1958 as a reaction to Sputnik and as a means for turning missiles into launch vehicles for America's civilian space efforts.

President Eisenhower opposed sending men to the Moon, but his successor, John F. Kennedy, made a lunar landing a national priority. Receiving virtual carte blanche in budget requests, NASA won the race for the United States, but victory was by no means an easy feat.

National insecurity, wounded national pride, infighting, political grandstanding, clandestine plots, and ruthless media frenzy were but a few of the things the United States had to overcome to bounce back from the blow dealt to the nation by Sputnik.

CHAPTER ONE

Sputnik Night

The news was a bombshell.

—Richard N. Goodwin, *Remembering America*

On a fall Friday afternoon in 1957, five bells rang ominously on noisy teletype machines in newsrooms across Washington, D.C., as a news wire brought word of Sputnik's launch.

LONDON, OCT. 4 (AP)—MOSCOW RADIO SAID TONIGHT THAT THE SOVIET UNION HAS LAUNCHED AN EARTH SATELLITE.

The news flash displaced several stories in the works: the tense racial situation at Central High School in Little Rock, Arkansas; the Milwaukee Braves–New York Yankees World Series; and a widespread flu epidemic. Jimmy Hoffa had been elected head of the Teamsters earlier in the day by a vote of 1,208 to 453. Yom Kippur was beginning at sundown, and the television series *Leave It to Beaver* would premier later in the evening on the CBS television network.

Details about the satellite were slow in coming, while information on the launch vehicle, or booster, that put Sputnik into orbit would not be known in the West for years. What was known in the first hours was that the Soviet Union had launched the first artificial satellite to orbit the Earth. It was about twice the size of a basketball, weighed only 184 pounds, and took approximately ninety-six minutes to orbit the Earth on an elliptical path.

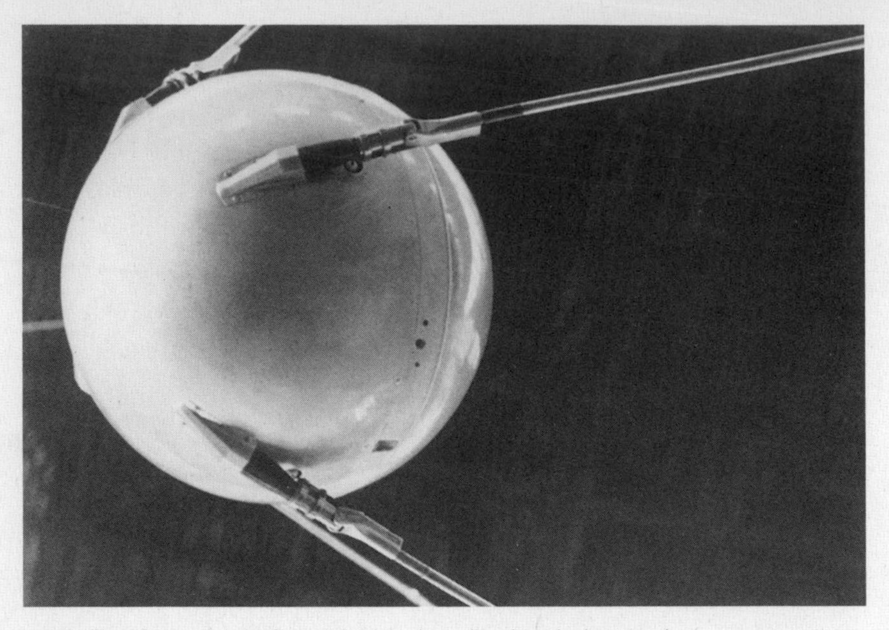

Replica of Sputnik I, the first artificial Earth satellite, launched on October 4, 1957. (NASA)

Shortly after 6 P.M. the news reached an international group of fifty or so scientists, many of whom were Russians and Americans, attending a party in the grand ballroom on the second floor of the Soviet Embassy in Washington. The scientists were participants in the International Geophysical Year (IGY), a grand sixty-seven-nation effort to unlock the secrets of the physical world. Officially deemed the "greatest scientific research program ever undertaken," the IGY involved more than 5,000 scientists in the effort to find out as much as they could about the Earth, the Sun, and outer space during the "year." (The IGY actually ran for eighteen months, from July 1, 1957, to the end of 1958, a period when there was maximum activity in solar flares.) Millions of facts would be collected, and major questions—such as whether or not the Earth's climate was changing—were to be investigated.

Earlier, the United States and the Union of Soviet Socialist Republics had announced to the world on separate occasions that each would put a small Earth-circling satellite into orbit as part of its contribution to the IGY. The Americans and much of the rest of the Western world had paid little or no attention to the Russians' plan but were eagerly looking forward to the launch of the first U.S. satellite.

Had it been on schedule, the Vanguard, the U.S. satellite, would have been launched November 1957. (Its anticipated timetable might have actually spurred on the Soviets.) However, Vanguard was eight to nine months behind schedule. There was no problem with the design of the satellite itself, but there were real problems with each of the three rockets needed to get it into orbit. The first stage lacked sufficient thrust, the second had to be redesigned, and the third was too heavy. Despite published reports alluding to slight delays, the American public perceived the program as moving along quite nicely. The Vanguard was now due to be launched in the spring of 1958, right in the middle of the IGY.

Vanguard was being built by the U.S. Navy, which had begun a massive publicity campaign to promote the satellite. By mid-1957, several books about Vanguard were already in the stores, and there were hundreds of feature articles about it in magazines. In May of 1957, a new edition of a popular book for hobbyists, *Discover the Stars,* was published with the image of Vanguard on the cover and detailed plans for building a model of the satellite. The book claimed that the Space Age would begin in early 1958 with a Vanguard launch from Banana River, Florida, also known as Cape Canaveral. *National Geographic* magazine referred to the planned Vanguard as "history's first artificial earth-circling satellite" (in February 1956) and as the "first true space vehicle" (in March of the same year). Martin Caidin noted in *Overture to Space* that "Vanguard had become a household word. . . . Scientists had given speeches and lectures on the miracle we were about to bring to the world. Artificial satellites had become synonymous with American genius, technology, engineering, science, and leadership."

"Everyone knew in 1957 that space exploration was the next item on the scientific and technological agenda, and almost everybody assumed that the United States would lead the way as usual," John Brooks wrote in *The Great Leap.* In fact, Americans were so complacent that they weren't even prepared to monitor other satellites. Therefore, on "Sputnik night" the Russian satellite twice passed within easy detection range of the United States before anyone in authority knew of its existence from the Associated Press report out of London.

Four days before the launch of Sputnik, the Comité Spécial de l'Année Géophysique Internationale (CSAGI), an international scientific IGY organization, opened a six-day conference at the National Academy of Sciences in Washington focusing on rocket and satellite research for the IGY.

Scientists from the United States, the Soviet Union, and five other nations met to discuss their individual national plans and to develop protocols for sharing scientific data and findings. However, the conference was abuzz because of a comment made by Sergei M. Poloskov, member of the Soviet delegation, at the opening session on Monday, September 30.

Poloskov's presentation was titled "Sputnik"—he pronounced it *spootnick*—the Russian word for "traveling companion of the Earth" and the name Russia had chosen for the satellite it was preparing to launch. Although earlier talk of a Soviet satellite had been dismissed, Poloskov's audience took notice when he used an expression that was translated into English as "now, on the eve of the first artificial Earth satellite." He announced that the transmitters in the projected Soviet satellite would broadcast alternately on frequencies of 20 and 40 megacycles. (A year earlier, the international ruling body for the IGY had stipulated a frequency of 108 megacycles as standard for all IGY satellites.) Speaking for the United States, Homer E. Newell pointed out to the Russian scientist that Project Vanguard's radio tracking stations, which were going on line the very next day, were set up to receive signals on the IGY-established frequencies. Since adapting the American equipment to receive the Soviet signals would require time and money, Newell asked Poloskov to say when his country hoped to put that first satellite in orbit.

The deftness with which Poloskov sidestepped Newell's question, along with similar questions from other delegates, produced such a roar of laughter that the sober Russian scientist himself finally and reluctantly joined in. All he would say was that when the Soviet satellite materialized, he hoped the Vanguard tracking stations would be able to collect the data it transmitted and send them to Moscow.

On October 4, Walter Sullivan of the *New York Times* was at the Soviet Embassy party when he received a phone call from his Washington editor. As quickly as possible he found Richard Porter, a member of the American IGY committee and chairman of its technical panel, and whispered, "It's up!" Sullivan had been scooped by events, for he had just filed a story with the *Times* for the next day that said the Russian satellite could go up at any moment. Although Porter had been convinced for days that a Soviet launching was imminent, he was still surprised that it had come so quickly and while the Russian scientists were still in town. He passed the information to Lloyd V. Berkner, an American physicist who was head of the Brookhaven National Laboratory and in charge of the American IGY program. Berkner clapped his hands, called for silence, and announced: "I've just been informed by the *New York Times* that a Russian satellite is in orbit at an elevation of 900 kilometers. I wish to congratulate our Soviet colleagues on their achievement."

The scientists and engineers assembled at the embassy party were thrilled. Cheers rang out. Within minutes, one of the most impenetrable buildings in Washington was putting out the welcome mat to reporters. The *Washington Daily News* later called it a veritable "open house." Vodka flowed as more news was given out about the satellite. The Americans offered their congratulations, and Berkner proposed a toast, while the Soviet scientists doled out proud quotes. Joseph Kaplan, chairman of the U.S. program for the IGY, called it "fantastic."

Someone brought out a shortwave radio, and soon a beeping noise filled the room. A Russian scientist, Anatoli Blagonravov, confirmed it was Sputnik. "That is the voice," he said dramatically. "I recognize it." John Townsend Jr., one of the scientists at the party, recalled watching Blagonravov: "I knew him quite well, and I could tell that he was a little surprised and quite proud. My reaction was 'Damn!' "

And so an abstraction now had a voice. It also had a name—Sputnik.

Many of those at the party adjourned to the Soviet Embassy's rooftop, attempting to view Sputnik with the naked eye. Several of the American scientists drifted over to the American IGY headquarters in Washington, where they began speculating on what impact the satellite would have. They feared that the American people would be disappointed.

It also dawned on them that they had better start tracking the satellite's orbit. They got in touch with the American Radio Relay League in West Hartford, Connecticut, asking its 70,000 members—all "ham" radio operators—to lend a hand and help track the Sputnik. In less than twenty-four hours, reports on the satellite were coming back to the National Science Foundation, where a temporary control room had been established. Eventually, these hams and other amateur and professional trackers would consider themselves part of a great international fellowship known as ROOSCH, or the Royal Order of Sputnik Chasers.

Listening to Sputnik at the Crimean Astrophysical Observatory of the Soviet Academy of Sciences. (Author's collection)

Huntsville Reacts

On the same evening, at about the same time, another cocktail party was going on in Huntsville, Alabama, at the Redstone Arsenal, where the Army Ballistic Missile Agency (ABMA) was working on Jupiter C, a powerful guided missile. The party was staged in honor of Neil McElroy, the visiting newly designated secretary of defense who was on an orientation tour before being sworn in. He was about to replace Charles E. Wilson, who was intensely disliked by the Huntsville missilemen for his lack of imagination and interest in their work. Wilson's greatest sin, they believed, was that he had given responsibility for long-range missiles to the Air Force and left the Army with a table scrap: missiles with a range of less than 200 miles. McElroy was in Huntsville to look at the Army's missile work and was accompanied by a large entourage from Washington, including the secretary of the Army and his chief of staff of the Army.

Hosting McElroy's group at the arsenal were Major General John B. Medaris, the Army's top missile commander and head of ABMA, and Wernher von Braun, the German rocket engineer who had headed the team that developed the V-2 rocket for Nazi Germany and now worked as the top missile scientist for the U.S. Army. These two, along with Lieutenant General James M. Gavin, the chief of research and development of the Army who had just arrived from Washington, had a major agenda: to launch and orbit their

Neil McElroy, incoming secretary of defense, arrives at Huntsville and gives an impromptu news conference earlier on the same day that Sputnik was launched. (U.S. Army, courtesy of Medaris Collection, Florida Institute of Technology)

own satellite powered by the Jupiter C. They had been pushing long and hard for this. As early as 1954, von Braun had tried to get permission to launch the Army satellite but was turned down. Even with the help of Generals Gavin and Medaris, Project Orbiter—as it had come to be called—had been turned down repeatedly by President Eisenhower, outgoing secretary of defense Wilson, and a group empowered to select America's IGY satellite.

McElroy and company arrived around noon and were briefed by von Braun, who once again made his pitch for an Army satellite to go into space. Medaris felt that the mission of his entire organization was to give the group from Washington the full and complete argument for the Army going into space.

The predinner "stag" cocktail party was in full swing, with McElroy, Medaris, and von Braun engaged in small talk, when Gordon Harris, the public affairs officer at the base, burst into the room, broke into the conversation, and said: "General, it has just been announced over the radio that the Russians have put up a successful satellite. It's broadcasting signals on a common frequency, and at least one of our local 'hams' has been listening to it."

There was an instant of stunned silence. General Gavin and others looked shaken. Then, as Medaris recalled later in his memoir, von Braun "started to talk as if he had suddenly been vaccinated with a Victrola needle. In his driving urgency to unburden his feelings, the words tumbled over one another. 'We knew they were going to do it! Vanguard will never make it. We have the hardware on the shelf. For God's sake, turn us loose and let us do something. We can put up a satellite in sixty days, Mr. McElroy! Just give us a green light and sixty days!' "

At dinner, McElroy was seated between Medaris and von Braun. There was a running fire of press updates on the Russian satellite, including the fact that it could now be heard on a radio at the base. Medaris did his best to sell McElroy on the idea of giving the Army the job of responding to Sputnik. Then Medaris dropped a bombshell. He said that more than a year earlier a Jupiter C designated Missile 27 would have put the nose of a rocket in orbit without question during a test "if we had used a loaded fourth stage."

Later, when everyone else had left, Medaris and von Braun lingered. They were angry and frustrated that the nation had been outmaneuvered, but were also "jubilant" because they assumed they would now be allowed to get their own satellite off the ground. The next morning they would use their brightest young officers to beg McElroy to let them get off the bench and into the game to score a touchdown for the West, America, and the U.S. Army. They knew they would have to make one hell of a sales pitch to convince their new boss. Fortunately, the only point on which Medaris and von Braun disagreed was that Medaris thought it would take ninety days to launch a satellite rather than the sixty that von Braun had promised over drinks. The fact that McElroy's visit coincided with the Sputnik launch created an optimal opportunity. Medaris later said they had been given "one of those little psychological breaks that happen only a couple times or once in a lifetime."

Early the next morning, von Braun and Medaris formally promised McElroy the first U.S. satellite in ninety days using the Jupiter C/Redstone rocket. "When you get back to Washington and all hell breaks loose," von Braun said, "tell them we've got the hardware down here to put up a satellite anytime."

After McElroy and his entourage left, Medaris told von Braun to get the mothballed Jupiter C rockets, starting with Missile 29, out of storage and "onto the floor"; the team went to work as if they already had a directive to proceed. It was a bold, risky move, which Medaris later recalled in his memoirs: "I was convinced that we would have final word inside of a week, and that week was too valuable to be lost. If we still did not get permission to go, I would have to find some way to bury the relatively small amount of money we would have spent in the meantime." He added, "I stuck my neck out."

Sputnik Makes a Lasting Impression

Dwight D. Eisenhower, an old Army man himself now in his second term as president, got the Sputnik news around 6:30 P.M. at his farm in Gettysburg, Pennsylvania. Before leaving Washington earlier in the day, he had been in meetings to discuss the federalization of the Arkansas National Guard and the use of federal troops in response to the crisis in Little Rock, which had been touched off when Governor Orval Faubus refused an order to desegregate the schools. Eisenhower was treating Faubus's defiance as an insurrection as well as a civil rights crisis. Later that evening, presidential press secretary James Hagerty advised news correspondents that "the Soviet satellite, of course, is of great scientific interest" but made a point of saying that the Russian announcement "did not come as any surprise; we have never thought of our program as in a race with the Soviets."

Word of Sputnik reached the headquarters of the Smithsonian Astrophysical Observatory in Kittridge Hall, in Cambridge, Massachusetts, at 6:15 P.M. The observatory's philharmonic orchestra was holding its first rehearsal of the season when Dr. J. Allen Hynek, the assistant director and ranking person on the premises, got the news in the form of a phone call from a Boston newspaper reporter, who asked, "Do you have any comments on the Russian satellite?" Within minutes, Smithsonian employees, scientists, and members of the media began to congregate at the observatory, which also was headquarters for the optical-tracking program set to follow the American Vanguard satellite. The observatory became the unofficial center for Sputnik information in the United States in the following hours and days. Within an hour Kittridge Hall was so ablaze with light from normally dark offices and camera crews that a woman living in the neighborhood reported that the building was on fire and a pumper and a hook-and-ladder went clanging to the scene.

As the evening progressed, Sputnik was heard by many people. At precisely 8:07 P.M., eastern daylight time, the signal was picked up by an RCA receiving station at Riverhead, New York, and relayed to the NBC radio studio in Manhattan. By this time Sputnik had already made three passes over the Western Hemisphere. Within moments, the sound of Sputnik was recorded for rebroadcast and could be heard everywhere there was a radio or television.

For years to come, Americans would recall where they were on Sputnik night. Senate majority leader Lyndon B. Johnson was at his ranch hosting

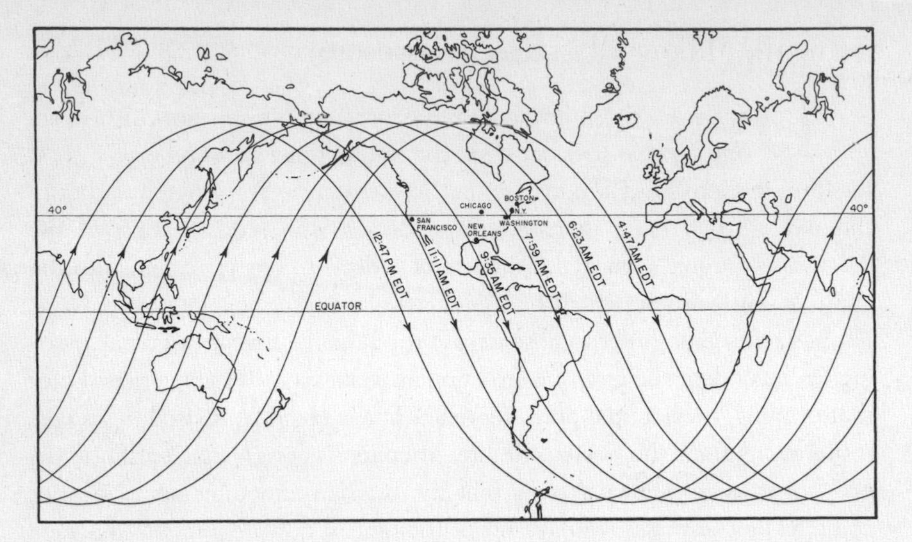

Orbits of USSR satellite released 10/8/57 by the U.S. Naval Research Laboratory. (U.S. Navy)

one of his trademark Texas barbecues when the news was announced. After dinner he, Mrs. Johnson, and their dinner guests took a long walk, as had become customary since his heart attack two years earlier. The once festive group was now silent as it looked skyward. "As we stood on the lonely country road that runs between our house and the Pedernales River," he later recalled, "I felt uneasy and apprehensive. In the open West, you learn to live with the sky. It is a part of your life. But now, somehow, in some new way, the sky seemed almost alien."*

Deeply moved by the event while also realizing it was a great political opportunity, Johnson immediately swung into action. He phoned his Senate colleagues of both political parties to get their support for investigative hearings on missiles and space.

Doris Kearns Goodwin, a former Johnson aide and now a presidential historian, recalls her own initiation into the Space Age. A sophomore in high school when Sputnik went up, she was at her boyfriend's house when the news was broadcast. The two decided they would go out and try to see it. "We took a blanket," she confessed on the *Newshour with Jim Lehrer* on the thirtieth anniversary of Sputnik's launch, "and we went to a park nearby.

*This account appears in Johnson's memoir *The Vantage Point.* Some associates later claimed that he actually saw it fly over his ranch in Texas that night, but this was not so. Perhaps it was more disturbing and ominous that he did not see it.

And it was a very romantic setting, and we started to look for Sputnik. And then my boyfriend reached over and kissed me. . . . I didn't give Sputnik another thought."

John F. Kennedy, who was then a U.S. senator from Massachusetts, seems to have shown even less interest in Sputnik—at least in public. Kennedy was a frequent closing-time visitor to the men-only bar at Boston's Loch Ober Café, where Freddy Hamil was maître d' and bartender. Hamil was smitten by space and was a devotee of Wernher von Braun, who was already a household name by virtue of his television appearances on the *Wonderful World of Disney.* Immediately following the Sputnik launch, Hamil introduced the future president and his brother Robert to Charles Stark "Doc" Draper, an MIT professor and pioneer in rocket guidance. A timely late-night bar conversation ensued on the meaning of the Russian feat. Many years later, Draper told aerospace historian Eugene M. Emme that it turned into an argument, with John Kennedy insisting ironically that all rockets were a waste of money and their use in space even more so. In retelling this story, however, Emme added, "But then the Kennedys were known to pick arguments just for the education of it or for the entertainment."

Alan Shepard, who would be the first American in space, was at the Naval War College in Newport, Rhode Island, when Sputnik was launched. He said that when he saw it in the October sky days later, he knew intuitively that "this little rascal" would affect him directly and quickly. John Glenn, who would follow Shepard as the first American in orbit, was already an American hero at the time. Weeks earlier he had set a new transcontinental jet speed record. He immediately saw the euphoria of that feat fade. "Supersonic flight had been outdone as a yardstick for measuring military superiority," he would say later.*

Half a world away, German Titov was just about to graduate from the Soviet Air Force pilots school when he heard the news of Sputnik and his mind raced ahead. "Maybe man can fly in space someday," he said to himself, "maybe in 20 to 25 years." Less than four years later, on August 6, 1961,

*Astronaut Alan Bean was a Navy pilot on shore leave standing in line for the evening show at the Lido nightclub in Paris, France. "A guy came up with a newspaper and announced [the Sputnik launch]," he later recalled. "We looked at the headlines and were all amazed. Even though I didn't really know what they'd done exactly or what it meant, I was amazed that the Russians could do something like that before we could."

he became the second human to go into space and the first to spend more than twenty-four hours in orbit. Konstantin Feokistov, the scientist who would fly aboard the 1964 Voskhod-1 (the first three-man capsule), had a different reaction: "When word of Sputnik reached me, I was very proud to be Russian. The world would now respect our science."*

Daniel S. Goldin, who eventually became the ninth NASA administrator, was a freshman at City College of New York. The Saturday following the launch, he went into physics class, where his professor had written "Sputnik Is Watching You" on the blackboard. He instantly became a "space nut" and knew that he wanted to work on a space program. Ed Stone, the director of NASA's Jet Propulsion Laboratory from 1991 to 2001, recalls that as a graduate student at the University of Chicago he saw doors open to a whole new area of science and technology in the aftermath of Sputnik.

Soon after the launch, biologist Max Dellbrück was hosting a picnic. (Among his guests was the great physicist Richard Feynman.) He hooked up a jury-rigged receiver, dialed up the Sputnik signal, quieted the group by putting an index finger to his lips, and then grinned broadly—as if to signal to his colleagues that science was back in the saddle.

"My life changed right there and then," Ross Perot recalled in a 1997 interview. He thought, "This is just like Kitty Hawk, the world is forever changed and I am going to be part of that new world." Ralph Nader, then a third-year law student at Harvard Law School, told *Air & Space* magazine, "It hit the campus like a thunderbolt." "Psychopathic" is how Harold W. Ritchey, the solid fuel rocket pioneer, described his shocked reaction. "It took me three months to get over it."

In Barcelona, Spain, where the eighth International Astronautical Congress was in session, word of Sputnik was received late, after most of the delegates had gone to bed. British writer Sir Arthur C. Clarke, the visionary who had been writing about the coming of the Space Age for years, was awakened from a sound sleep by reporters asking for comments. He told them that Sputnik would have "colossal repercussions."

*Titov and Feokistov were interviewed on May 15, 1997. The occasion was the opening of the exhibit "Space Race" at the Smithsonian. One of the objects in this exhibit was Feokistov's flight suit, which had been bought in 1993 by Ross Perot at a Sotheby's auction. At the opening of the exhibit, Perot revealed that he had used surrogate buyers to obtain a huge collection of Russian space artifacts. "This is their heritage and somebody has to protect it and eventually give it back to them," said the billionaire. "They love me in Russia."

Observing Sputnik through telescopes at the astronomic post of the Novosibersk Institute of Geodasy, Aerial Photography and Cartography. (Author's collection)

On his way home from a Black Sea vacation, Nikita Khrushchev stopped in Kiev, where he awaited news of the launch. His son, Sergei, later recalled that at about 11 P.M. his father got word from the launch site that the satellite was in orbit and shortly thereafter heard its transmission on a shortwave radio. Deeply impressed with the feat, he could not fully understand its impact until he saw how the rest of the world, especially the West, reacted to it.

Sputnik night is recalled in many memoirs and recollections, and it is the rare writer who, recalling the night, does not admit to being overwhelmed by its historic importance. Coincidentally, writer James A. Michener was in flight on a military DC-3 from Guam to Tokyo when the plane ran into trouble and was forced to ditch in the Pacific. Everyone ended up in a large rubber raft. The group was rescued and flown to the Iwakuni base near Tokyo, where an excited reporter shouted, "Have you heard the news?" Michener, as spokesman for the group, answered, "Yes. We ditched in the middle of Pacific." The reporter shouted, "No! The Russians have sent a spaceship into orbit around the world." As Michener would later recall, "Within minutes we had forgotten our own adventure in the shadow of one so infinitely greater."

One of the lucky individuals who caught a glimpse of Sputnik was Saunders Kramer, cofounder of the American Astronautical Society. Kramer heard about it while working for the Lockheed Missiles and Space Company in Palo Alto, California. Then he listened to the beeping of Sputnik on his car radio on the way home. The next morning he got up at 4:30 and went out on his patio to look for the satellite with binoculars. In the October 1987 issue of *Space World* magazine, Kramer recalled that before he actually saw Sputnik, he thought, "What am I doing here, the only person crazy enough to be out here this early on a Saturday morning." But moments later, his neighbors all the way down the block were looking up and saying, "Do you see it? Do you see it?" for the next several minutes. And then, precisely at 5 A.M., out of the northwest sky, Sputnik appeared. Kramer raised his binoculars and saw the satellite when "suddenly a huge meteor slammed across the sky, leaving a trail of orange ash which lingered for several seconds. I obviously wasn't the only one who saw it because it made the front page of the *San Francisco Chronicle* the next morning. I'll never forget it."*

The Press Reacts

When the Sun came up in the United States the day after Sputnik's launch, the country experienced a sense of awe rather than panic. As would be true for many weeks to come, President Eisenhower and his top advisers reacted calmly. On that Saturday morning, and for the fifth time that week, the president of the United States played golf.† A *Newsweek* correspondent in Boston wrote in a memo that same morning that the "general reaction here indicates massive indifference," while another *Newsweek* writer wired his home office from Denver that there "is a

*Gordon Little, a British astronomer, made the first official sighting from a professional observatory in the United States. The sighting was at the Geophysical Institute, University of Alaska, Fairbanks, at 5:01 A.M. local time (11:01 EDT) on Sunday, October 6. Carla Helfferich, a science writer at the institute, in an article for the *Alaska Science Forum* on October 9, 1997, confirms this but then defers to T. Neil Davis's claim in his book *Alaska Science Nuggets* that it was one Dexter Stegemeyer who was the first Fairbanksan known to have seen Sputnik just clearing the western horizon, when he looked through the doorway of his open outhouse. Stegemeyer knew he'd seen something unusual, but realized what it was only after he heard a description of the satellite.

†In the days ahead, much was made of Ike's reaction. The *Birmingham News* (Alabama) ran the story under a separate headline, "Ike Plays Golf, Hears the News," and the *Nashville Tennessean* depicted the president as pooh-poohing Russian advances from the golf course.

vague feeling that we have stepped into a new era, but people aren't discussing it the way they are football and the Asiatic flu."

Polls taken within days of the launch showed that Americans were concerned—so concerned that almost every person surveyed was willing to see the national debt limit raised and forgo a proposed tax cut in order to get the United States moving in space.* A Gallup poll for *Newsweek* found that 50 percent of a sample taken in Washington and Chicago regarded Sputnik as a blow to U.S. prestige. Still, 60 percent said that America, not Russia, would make the next great scientific advance. A poll by the *Minneapolis Star and Tribune* found that 65 percent of Minnesotans thought the United States could send up a satellite within thirty days following the Russian success. In a quick survey conducted by the Opinion Research Corporation, 13 percent believed that America had fallen dangerously behind, 36 percent that it was behind but would catch up, and 46 percent that it was still at least abreast of Russia. Assistant Director J. Allen Hynek of the Smithsonian Astrophysical Observatory had the impression that Americans, on this fine autumn weekend, felt they had "lost the ball on [their] own 40-yard line but would still win the game."

The initial media reaction was diverse. The *New York Times* announced the event in an extremely rare three-row headline with much supplementary information. The *Milwaukee Sentinel* relegated the story to a small headline and short article on page three, while the front page of the paper proclaimed "Today We Make History"—because the city was hosting the World Series for the first time.

The only discord that Saturday morning was from Huntsville, where a scientist "asking that his name not be used" (this was almost certainly the media-savvy Wernher von Braun) told the Associated Press that he was "angry and distressed" because the Army could have had a satellite in orbit if it had been given the assignment in 1955. Medaris and von Braun had apparently decided that they would publicize their rage against Washington.

Over the weekend the news media, still collectively known as "the press" in those days, realized that Sputnik was a big, big story. They needed a means of reporting it, so they besieged scientific and military institutions in search of authoritative voices to provide datelines and interpret events.

*Public opinion at any time, of course, has filters and caveats attached. Historian Kim McQuaid pointed out that Sputnik opinion polls generally were limited to "the 5–7 percent of the adult population who are college graduated, white collared, 'professional,' or, in large part, upper middle to upper class."

The Smithsonian Astrophysical Observatory in Cambridge, Massachusetts, was flooded with reporters, who ended up staying for weeks, while other reporters latched on to the willing Army sources at Huntsville or the Army's Jet Propulsion Laboratory in Pasadena.

In England, the Jodrell Bank Observatory, the world's largest radio telescope—though not yet completed at the time—was thrust into the role of satellite central for the United Kingdom and western Europe. Sir Bernard Lovell of the observatory wrote in his 1968 memoir: "Throughout Saturday and Sunday a state of siege of newspaper and broadcasting personnel began to develop around my house and Jodrell." Within hours, the BBC crew alone outnumbered the staff at the observatory.

Some things were immediately clear to the legions of reporters and editorialists assigned to the story. First and foremost, the launch of Sputnik into orbit signaled the very moment when the Space Age began. Although the *London Daily Express* was the first to actually proclaim it in a headline—"The Space Age Is Here"—the term *Space Age* now cropped up everywhere. Writers with scant details on the satellite opted for Sunday "thumbsuckers"—journalistic slang for labored, reflective essays often written in lieu of hard news—about the dawning of a new age. Typically, these stories told readers born in the horse-and-buggy era that they could now claim to have made it to the Space Age.

Also, there was no question that the Soviets had scored a major scientific achievement, and there was no shortage of experts to attest to this. Sir Arthur C. Clarke, for example, said that the launch was "one of the greatest scientific advances in world history." Sir Bernard Lovell labeled it "absolutely stupendous, about the biggest thing that has happened in scientific history."

In those first hours of the Space Age, many writers gave Sputnik a special identity, which added to its luster and romance. It was declared to be nothing less than a new moon.* The headline writers loved the lunar label. "Made-in-U.S.S.R. 'Moon' Circles Earth; Space Era Advent Jolts Washington" was the first-day banner headline in the *Christian Science Monitor,* while "Russia Launches a Moon" appeared in the *London Daily Mail.*†

In the United States especially, newspapers were quick to draw their readers' attention to the fact that Sputnik was flying overhead. Saturday

*One newspaper, the tabloid *New York Daily News,* called it "Khrushey's Comet."

†When America's first satellite, Explorer 1, finally went up, it was America's moment to gloat. "The United States hung a moon over Russia tonight" was the lead filed by the UP's William Tucker.

morning's *Cleveland Plain Dealer* screamed in a two-line headline, "Satellite Fired by Russia; Circling US 15 Times a Day." Diagrams and maps showing the overflight path were a common sight in U.S. newspapers. The American press also conceded that the Soviet Union had won some points in what was, to use a Cold War cliché, "the war for men's minds." "Major Propaganda Victory Believed Scored by Russia" read a *Denver Post* Saturday headline, which was echoed in the simple Sunday *New York Times* headline "A Propaganda Triumph." The *Times* backed up its headline with an editorial terming the Soviet announcement "one of the world's greatest propaganda—as well as scientific—achievements."

By Sunday, the rest of the world had had a chance to react. The satellite was the lead story in every British paper, and all were awed by the Russian feat. The *Sunday Telegraph,* however, added that it believed that the United States soon would surpass the Soviet Union in space. The Chinese papers viewed Sputnik as proof that the Communist system was superior and had superior scientists. In Austria, a Communist newspaper, *Volkstimme,* commented with pride, "In contrast with the first step into the atomic age which began with 100,000 deaths and frightful destruction in Hiroshima and Nagasaki, mankind can rejoice without destruction on the . . . conquest of the cosmos by the human spirit." *Die Presse,* a non-Communist Austrian paper, asked who could be certain that "the satellite is intended not primarily for scientific purposes or the exploration of space but preparation of war on a planetary scale."

By the end of the weekend, the giddying effects of the event were wearing thin in the United States. The scientific, political, military, and media elite were no longer in a congratulatory mood. Nor, it seems, was the American public. For one thing, while Sputnik put a proud Soviet Union in the world spotlight, Americans were hard-pressed to find an upside to the story. Reporters hunting for a positive spin had to settle for the rather feeble notion that the United States, purportedly the most powerful scientific nation on Earth, was up to the challenge of tracking the first man-made object to leave the planet. The *New York Herald-Tribune* reassured its readers with the headline "U.S. Scientists Map Red Moon's Orbit."

The only other positive bit of news was a dispatch from Moscow reporting that the Russian people did not learn about the launch until after most Americans and much of the rest of the world had. This highlighted the fact that American society enjoyed a free flow of information, whereas Russian society did not.

A press quote proved to be prophetic. Rocket scientist and avocational science fiction writer G. Harry Stine was fired from the Glenn L. Martin Company, the prime contractor for the Vanguard satellite program, because he was quoted in a Denver newspaper interview on Saturday saying, "We have known in the rocket business for a long time that the Russians were pretty sharp. . . . We lost five years between 1945 and 1950 because nobody would listen to the rocket men. We have got to catch up those five years fast or we are dead." Stine later pointed out to an Associated Press reporter that the comments that cost him his job simply paraphrased what he had written in his book *Earth Satellites and the Race for Space Superiority,* published in paperback a month before the Sputnik launch.* Within days, Stine's comments would be echoed by many.

Meanwhile, Russian rocket scientists still in Washington for the IGY conference had become the instant darlings of the radio and television media. On Saturday, three of them appeared on the NBC-TV show *Youth Wants to Know.* Anatoli Blagonravov was asked if the satellite was a victory over the West. "We did not consider it necessary to compete in this field," he answered, "and we would be happy, no less than we are happy now, if we see the American satellite in space. We believe that our satellite, as well as the American satellite, could do it and serve science."

But before leaving town, the Russians did take one final poke at their American counterparts. On Saturday, in the auditorium of the U.S. National Academy of Sciences, Blagonravov was given the floor to speak about Sputnik. Homer E. Newell, who was in the audience, later recalled: "Understandable pride was evident in Blagonravov's bearing, but his words also bristled with barbs for his American listener. The speaker could not refrain from chiding the United States for talking so much about its satellite before having one in orbit, and commended to his listeners the Soviet approach of doing something first and then talking about it."

Newell and the other Americans felt that Blagonravov's "ungracious" comments missed the point of the IGY, which was to talk about projects before, during, and after so that others could share information.

On Sunday, the Russians were making more news in the United States as the country learned of a press release issued by Tass, the official Soviet news agency, which reported: "During the International Geophysical Year the Soviet Union proposes launching several more artificial earth satellites.

*The teaser at the front of the book read, "For the first time since the dawn of history the Earth is going to have more than one moon. This is due to happen within the next few months—or it may have already happened even at the time you are reading this."

These subsequent satellites will be larger and heavier and they will be used to carry out programs of scientific research." Tass ended its release with this line: "Artificial earth satellites will pave the way to interplanetary travel, and apparently our contemporaries will witness how the freed and conscientious labor of the people of the new socialist society makes the most daring dreams of mankind a reality."

The U.S. public immediately began to learn more about rockets and satellites, including the fact that a Russian was the first person to prove the theory of spaceflight more than fifty years before Sputnik. News services picked up a story from the October 5 *Pravda* that said, "As early as the end of the nineteenth century the possibility of realizing cosmic flights by means of rockets was first scientifically substantiated in Russia by the works of the outstanding Russian scientist Konstantin E. Tsiolkovsky." Russian rocket literature was largely unknown in the West, although any Russian could read the work of Western rocket experts.* And now the heirs to Tsiolkovsky had put an aluminum alloy sphere in orbit. Many wondered what held it up. Schoolteachers, reporters, and editorialists found themselves dipping into the theories and laws of Sir Isaac Newton, who was the first one to explain—almost 300 years earlier—how a satellite could work.

*Willy Ley, the pioneering German-born space writer and protégé of Hermann Oberth, wrote in an introduction to Ari Shternfeld's *Soviet Space Science*, "Thus while it was easy for the Russians to acquaint themselves with Western ideas and work in this field, the English-speaking peoples almost completely overlooked what the Russians were doing." Ley concluded, "Sputnik I proved not only the validity of Newton's laws but also the dangers of insularity."

CHAPTER TWO

Gravity Fighters

The earth is the cradle of the mind, but one cannot live forever in a cradle. . . . From the moment of using rocket devices a new, great era will begin in astronomy; the epoch of the more intensive study of the firmament.

—Konstantin Eduardovich Tsiolkovsky, Russian rocket pioneer

Two individuals, separated by as many centuries, envisioned the possibility of an artificial satellite orbiting the Earth. The first was English mathematician and cleric Sir Isaac Newton, who, in 1687, mentioned the idea as a theoretical possibility as he explored the laws of gravity. The second was Konstantin Eduardovich Tsiolkovsky; in 1898 he created the first formula for propelling an object into orbit.

Newton realized that the same force of gravity pulling an apple to the ground also reached into the vacuum of space, and that it was the Earth's gravity that held the Moon in orbit. Gravity extended beyond the Earth and Moon and through the universe. His law of universal gravitation explained why the planets continue to revolve around the Sun instead of flying off into space.

Despite appearances to the contrary, moons and orbiting satellites do not defy gravity; in fact they are falling toward the Earth all the time. The reason a satellite doesn't plummet back to Earth is that the satellite is moving extremely fast and at a great height. As the satellite gradually falls toward Earth, the Earth's surface curves away from it; therefore, the satellite gets closer very, very slowly. If an apple is thrown toward the horizon, the apple falls to the ground within a few yards, following a curved path. Newton believed if the apple could be thrown fast enough, the downward curve of the apple's path would match the curve of the Earth's surface. In a sense the Earth would be a moving target, continually dropping away before the apple fell far enough to hit the ground. Sputnik stayed up for the same reason the Moon does—the pull of the Earth's gravity and the velocity of the satellite were perfectly balanced. Like the Moon, Sputnik had too much velocity to fall to Earth but not enough to break away from the Earth's gravity. This is the same principle by which the Earth and other planets are held in orbit by the strong gravity of the Sun. The trick to getting Sputnik into space was firing a rocket that could defy gravity and put the satellite into orbit at just the right speed.*

Newton actually described how an artificial satellite could be launched. He pictured a powerful cannon on top of a high mountain firing shots parallel to the ground. Each time the big gun was fired, more gunpowder was used and the shot went farther before it struck the Earth. Newton's theory held that the shot could in fact go so fast that it could curve around the Earth and come back to the mountaintop.

Not quite six months after Sputnik I, President Eisenhower gave the American people a space primer prepared by his Science Advisory Committee, chaired by James R. Killian Jr., president of MIT. *Introduction to Outer Space* explained that to escape the bounds of Earth the rocket must achieve a velocity of 25,000 miles per hour. "Although the basic laws governing satellites and space flight have been well known to scientists ever since Newton, they may seem a little puzzling and unreal to many of us. Our children, however, will understand them quite well."

"The Brick Moon"

The idea of an artificial moon in Earth orbit remained a Newtonian abstraction until 1870, when an unusual short story titled "The Brick Moon" was published in the *Atlantic Monthly* magazine. It was written by Edward Everett Hale, a Unitarian minister, chaplain to the U.S. Sen-

*In those first few days after the launching of Sputnik, many Americans and most of the press corps were doing their utmost to figure out why it just didn't fall down. A straight shot out to the Moon or Mars would have been much more understandable than this unnamed hunk of metal flying overhead at predictable times.

ate from 1902 to 1909, and an imaginative writer. He was known to generations of Americans as the author of "A Man Without a Country," a short story about Philip Nolan, a fictional conspirator in Aaron Burr's 1803 attempt to create a separate nation in the American West.*

The premise of "The Brick Moon" was that a small but clearly visible body orbiting the Earth would be invaluable to navigators, who, with the aid of a sextant and the proper tables, could use it to determine exact locations. Anyone could look up from anywhere on Earth and calculate his or her longitude relative to the position of the orbiting moon, which would move perpendicular to the equator. Hale likened the path of his moon to "a ring like Saturn's which stretched around the world, above Greenwich and the meridian of Greenwich, . . . anyone who wanted to measure his longitude or distance from Greenwich would look out of his window and see how high this ring was above his horizon."

In the story, a band of young, altruistic students in the classroom of a Boston college come up with the idea and spend the next thirty years raising the money to build the Brick Moon. The funds come from their own earnings, saved specifically for the project, and from public subscriptions. The moon is made of brick, rather than iron, to withstand the friction of passing through the atmosphere at high speeds. (Brick controls heat in much the same way that today's space shuttle uses ceramic tiles for protection.) The large, hollow moon is divided into seven spacious living chambers and storage rooms for food and other goods.

In Hale's story complications ensue and the launch is threatened, but the Brick Moon finally makes it into orbit.† The story ends happily with the thirty-seven men, women, and children aboard living in a state of perfect harmony while saving lives on Earth. There is plenty of air, food, and friends within the Brick Moon, where it rains regularly, providing a steady source of drinking water. The near-tropical climate helps to form a soil, making it possible to grow crops of palms, breadfruit, bananas, oats, maize, rice, and

*Hale was noted for his quick wit. For example: "Do you pray for the Senators, Dr. Hale?" asked a visitor to the Capitol. "No," snorted the newly appointed Senate chaplain, "I look at the Senators and pray for the country!" (Quoted by Van Wyck Brooks in *New England: Indian Summer*.)

†The moon is being built at a remote location in the mountains of western Maine, and when completed, it is to be rolled down a gigantic groove cut in the earth and shot into space as it comes into contact with two enormous water-powered flywheels turning in opposite directions. It will then be "snapped upward, as a drop of water from a grindstone." During construction in Maine, the workmen, engineers, and their families all

wheat, which are harvested up to ten times a year. Gifts from Earth are wrapped in layers of woolen carpet and shot at the Brick Moon, with the layers burning and disintegrating on the voyage. Some of the gifts are lost, some arrive intact, and some, alas, become moons of the satellite.

Hale's fantasy created a momentary sensation and was afforded a modicum of the same kind of attention given to Jules Verne's *From the Earth to the Moon*, published four years earlier. President Ulysses S. Grant was widely reported to have made the apocryphal remark that the Brick Moon was "the biggest thing since Creation, save for the invention of bourbon whiskey and the Havana cigar."

The fictional touch that makes the story especially prophetic is that there is communication from the Brick Moon to the Earth below. The inhabitants keep in touch with friends and family on Earth by making long and short jumps, producing messages in Morse code that are visible through a telescope. This is the first time that a crude form of telemetry from space—the transmission of information without wire, cable, or physical delivery—is envisioned. Less than one hundred years later, when America would undertake a more practicable scheme for a manned satellite in Project Mercury, the telemetry would be electronic.

Sir Arthur C. Clarke, himself a Space Age prophet who was the first to imagine communications satellites, once called "The Brick Moon" a more amazing anticipation "because it contained the very first proposal for an artificial satellite—as a navigational aid! Shades of GPS." GPS is shorthand for Global Positioning System, whose predecessor came into being shortly after the first Sputnik went up. Using the radio signals from satellites in precisely defined orbits, scientists at Johns Hopkins University's Applied Physics Laboratory discovered that a receiver could be used to determine exactly where the receiver was on the face of the Earth by measuring its position relative to the satellites. In this regard, Sputnik helped realize, at once, both the search for longitudinal perfection and the idealism behind the Brick Moon. Today GPS tracks animal migrations, monitors glaciers, guides motorists, finds 911 emergency callers, and pinpoints the location of ships at sea.

---

moved inside the moon, where it is warmer and much more comfortable than in the log cabins that have been provided for them. An Earth tremor causes the satellite to break loose, roll down the groove, and shoot 5,000 miles into Earth orbit—all in a matter of seconds. For a while it seems to be lost in space, but then a memorandum appears in the journal *Astronomische Nachrichten* that the Brick Moon has been seen by Professor Karl Zitta of Breslau, who names it Phoebe.

The Early History of Rocketry

Neither Newton's satellite nor Hale's vision for an artificial moon could be realized until there were energy sources powerful enough to put an artificial object in space. Newton's concept of a big gun was not sufficient. Rockets were the answer, and their power depended on Newton's third law of motion, which states: "For every action there is an equal and opposite reaction, and the two are along the same straight line." For example, if someone seated in a canoe fires a machine gun, the gun's recoil will cause the boat to move in the direction opposite that of the gunfire. This forms the basis of modern rocketry.

Invented by the Chinese around A.D. 1200, the earliest rocket was a tube, capped at one end and containing black powder. The other end of the tube was left open, and the tube was attached to a long stick. When the powder was ignited, the rapid burning of the powder produced fire, smoke, and gas, which escaped out the open end and produced a powerful thrust. The stick acted as a simple guidance system that kept the rocket headed in one general direction as it flew through the air. These rudimentary rockets did not travel very far, were inaccurate, and were dangerous to fire, but they had a range that was greater than a spear or an arrow.

The first true practical application of rockets was in 1232, during the battle of Kai-Keng, when the Chinese repelled the Mongol invaders with a barrage of "arrows of flying fire," which were probably more effective as weapons of terror than of destruction, although they were capable of setting fires. Rocket use made its way from China to India and then on to the Middle East and Europe. In Europe rockets were employed to support attacks of fortified cities in the late thirteenth century. They continued to be improved, especially after the invention of gunpowder, which was a much better propellant than black powder.

From the thirteenth through the fifteenth centuries there were periodic advances in rocketry. The English monk Roger Bacon worked on improved forms of gunpowder that greatly increased the range of rockets. Frenchman Jean Froissart found that more accurate flights could be achieved by launching rockets through tubes, an idea that was the forerunner of the modern bazooka. An Italian, Joanes de Fontana, designed a number of rocket weapons, including a surface-running rocket-powered torpedo intended to set enemy ships on fire.

By the sixteenth century rockets fell out of favor as weapons of war, through they were still used for fireworks displays. A German pyrotechnician,

Depiction of an Indian rocketeer, dated 1792, from the Library of the India Office in London. (NASA History Office)

Johann Schmidlap, invented the "steprocket," a multistage vehicle for lifting fireworks to higher altitudes. When a large first-stage rocket burned out, a smaller one continued to a higher altitude and exploded, showering the sky with glowing cinders. The steprocket is basic to all modern spacecraft.

The desire for new weapons eventually brought back the idea of using rockets in warfare to set fires, terrorize one's enemy, and inflict casualties. Indian rocket barrages were used successfully against the British in 1792 and again in 1799. The Indian rocket weighed as much as twelve pounds, was encased in an iron tube, and was stabilized by a ten-foot bamboo pole—a deadly projectile in itself. This rocket piqued the interest of an artillery expert, Colonel William Congreve, of the Royal Laboratory at Woolwich. Congreve set out to design rockets for use by the British military.

On October 8, 1806, the British launched 200 Congreve rockets against the French at Boulogne, the first military rockets ever used in Europe. In 1813 a barrage of Congreve rockets burned out Napoleon's supplies and forced the surrender of Danzig. The United States was also at war with the British at this time, giving Congreve's missiles their North American debut.

These rockets were potent weapons in Britain's naval and land arsenal. The thirty-two-pound Congreve warhead consisted of a hollow, cylindrical, iron body forty-two inches long and four inches in diameter. Within the rocket tube was the black powder propelling the charge, and attached securely to the metal body was a fifteen-foot-long wooden stabilizer. Since his rocket could be launched without damaging its launch site, Congreve was able to design very light and mobile devices to launch the rocket on land or at sea. The Congreve rocket had a maximum range of 3,000 yards and could be fitted with interchangeable warheads: incendiary, shot, and explosive. The explosive rocket was used in the September 1814 attack on Fort McHenry in Baltimore, while the incendiary version was used in shore bombardments along the Chesapeake Bay, setting ships and barges afire, and according to one account, burning the White House.

The rocket attack on Baltimore is etched in America's collective memory. For two phrases in the national anthem—"the rocket's red glare, the bombs bursting in air"—refer to the use of the Congreve rockets during the shelling of the fort. The actual rocket bombardment lasted about twenty-five hours, killing four Americans and wounding twenty-four. A metal rocket casing on display today in the Visitor Center at Fort McHenry was fired at the home of Henry Waller on Maryland's eastern shore, in Kent County. After the war, in an attempt to recoup his losses, Waller engaged Francis Scott Key, the young lawyer who wrote the national anthem, to take his case.*

Even with Congreve's work, the accuracy of rockets still had not improved much since the early days. The devastating nature of the British war rockets had nothing to do with their accuracy or power but with their sheer numbers. For example, during the 1807 siege of Copenhagen by the Royal Navy, 40,000 rockets were fired at the city, burning a large portion of it to the ground.

The War of 1812 inspired American experimentation with rockets. On December 4, 1846, a brigade of rocketeers was authorized to accompany Major General Winfield Scott's expedition against Mexico. The U.S. Army's first battalion of rocketeers, about 150 men armed with about fifty rockets, was used on March 24, 1847, against Mexican forces at the siege of Veracruz and again on April 8 when the battalion moved inland and about thirty rockets were fired during the battle for Telegraph Hill.† The battalion later used

*When Sputnik was regarded as the "second" foreign rocket attack on America, some editorialists put a twist on the phrase "rocket's red glare," changing it to "red rocket's glare."

†The job of placing those rockets in firing position fell to Captain Robert E. Lee, who would later command the Confederate Army of Northern Virginia in the Civil War.

rockets in the capture of the fortress of Chapultepec, which, in turn, forced the surrender of Mexico City.

By the middle of the nineteenth century, the rocket again disappeared as a weapon of war as the range and accuracy of artillery improved. All over the world, rocket researchers continued experimenting with ways to improve rocket accuracy, but there were few successes. One of them was a technique called spin stabilization (developed by an Englishman, William Hale), by which escaping exhaust gases struck small vanes at the bottom of the rocket, causing it to spin much as a bullet does in flight.

Through the nineteenth century rockets had a relatively short range, could not carry a heavy payload, and were unguided after launch. They saw limited use in combat, with only occasional moments of triumph. Their first recorded use in the American Civil War came in 1862, when Major General Jeb Stuart's Confederate cavalry fired rockets at Major General George B. McClellan's Union troops at Harrison's Landing, Virginia. The rockets had little, if any, impact other than as an expensive fireworks display. Later the same year, an attempt was made by the Union Army's New York Rocket Battalion—160 men under the command of British-born Major Thomas W. Lion—to use rockets against Confederates defending Richmond and Yorktown, Virginia. When ignited, the rockets skittered wildly across the ground, passing between the legs of a number of mules.

A few individuals could see rockets as something other than weapons and were finding new applications for them. Inventor John Dannet created a lifesaving rocket that was used to fire a line from the shore to vessels in distress. Others created rocket-powered, explosive-tipped harpoons for whaling. Fanciful fiction writers saw rockets as vessels for space travel. Jules Verne, one of the leading writers of his time as well as one of the founders of science fiction, wrote of a Moon rocket fired from a 900-foot-long cannon in his fantasy adventure *From the Earth to the Moon.** H. G. Wells, on the other hand, wrote novels in which the future was filled with technology that was often disastrous. Mainly through his *War of the Worlds,* he suggested the possibility of a hostile universe intent on the destruction of humanity.

Verne and Wells inspired the early space pioneers. "Both were well aware of the scientific underpinnings for spaceflight and their speculations reflected reasonably well what was known at the time about its problems and the nature of other worlds," according to space historian Roger

*Verne's rocket was fired from Florida west of and at the same latitude as Cape Canaveral.

D. Launius. "They incorporated into their novels a much more sophisticated understanding of the realities of space than had been available before. Their space vehicles became enclosed capsules powered by electricity, and they possessed some aerodynamic soundness. Most of Wells's and Verne's concepts stood up under some (but not too rigorous) scientific scrutiny."

In World War I the Allies used rockets for signaling, battlefield illumination, and the laying of smoke screens both on land and at sea. The French developed Le Prieur rockets, named after naval lieutenant Y.P.G. Le Prieur, who suggested them; they were fired with limited success from aircraft or the ground at vulnerable German zeppelins and balloons.

A handful of American inventors and engineers became intrigued with the notion of putting a weapon into the air with some kind of guidance other than an aviator. This was the departure that ultimately led to the creation of guided missiles and unmanned spacecraft. American technology of World War I included several unmanned flying vehicles such as the Kettering "Bug" and Sperry "Aerial Torpedo" used in 1917 and 1918. They were winged carts-on-wheels with engines that somehow managed to lift them into the air after a fast start on a pair of rails. Though they flew, they were unsophisticated and were never used as accurate, winged bombers. The day of these unwieldy monsters ended with the Armistice. In the next world war, rockets and armed pilotless missiles would return with a vengeance.

The principal weakness of rockets during the Great War and for the next two decades was the type of fuel used to power them. The black powder, gunpowder, and other dry explosive concoctions of the previous seven centuries were simply too weak and unreliable. Experiments were under way in the United States and across Europe to develop a more powerful and efficient liquid-propelled rocket.

Russian Visionary

The rockets that put Sputnik and all subsequent spacecraft into orbit had origins dating back to the end of the nineteenth century. Three individuals—one in the Soviet Union, one in the United States, and one in Germany—created modern rocketry. The first was the Russian Konstantin Eduardovich Tsiolkovsky. Born in 1857, Tsiolkovsky was the son of a woodcutter who had become deaf at an early age from scarlet fever. He was

Russian space pioneer Konstantin Eduardovich Tsiolkovsky. (Sovfoto)

a largely self-taught scholar, who sometimes lived almost entirely on black bread so that he could afford books. Tsiolkovsky became a mathematics teacher in the isolated town of Borovsky in the Kaluga province, and as early as 1883, he wrote his first articles on space, explaining how it would be possible for a rocket to fly in the vacuum of space. This was at a time when most people believed it was not even possible for a vehicle to fly in the air!

In 1895, Tsiolkovsky published *Dream of the Earth and Sky,* in which he wrote that an artificial Earth satellite might be possible. The notion of an Earth-circling "moon" previously had been limited to fiction, such as Hale's Brick Moon.

In 1898, five years before the Wright brothers flew at Kitty Hawk, Tsiolkovsky created a formula specifying what it would take to escape the Earth's gravity and described how a rocket—he called it a "reaction machine"—was the way to accomplish the feat. In 1903 he began publishing parts of his book *The Investigation of Space by Means of Reactive Devices.* In it he proved mathematically the feasibility of using Newton's reactive force to eject a vehicle into space above the pull of the Earth's gravity. His basic equation—the first theoretical proof of spaceflight—is still known as the "Tsiolkovsky equation." This formula was given scant attention inside czarist Russia and none whatsoever outside.

Tsiolkovsky was a theorist using few experiments and no known working rockets, yet he had many innovative and prophetic ideas. He was the first to recommend the use of liquid propellants because they would perform better and would be easier to control than solid propellants.* He suggested controlling a rocket's flight by inserting rudders in the exhaust or by tilting the exhaust nozzle; he imagined valves as a way of controlling the flow of liquid propellants. He even thought about putting humans into space. His cabin designs included life support systems for the absorption of carbon dioxide. He was the first to propose reclining the crew with their backs to the engines during acceleration. He envisioned the assembly of manned orbiting space stations sustained with food and oxygen supplied by vegetation growing aboard the craft.

Tsiolkovsky lived in poverty until the 1917 Russian Revolution, when the Bolsheviks embraced him for his views on human social evolution. He believed that humans, because of their ability to build and use tools, would conquer gravity, enter space, and spread across the cosmos, building a perfect society in their wake. This fit so well with Leninism that the Bolsheviks gave Tsiolkovsky the title "Father of Cosmonautics." Lenin personally endorsed a liberal pension that would allow him to work freely at his research for the rest of his life. His work inspired others, and he was the cofounder of the Russian Society for the Study of Interplanetary Communications. Rocketry and revolution were made for each other.

In 1926, Tsiolkovsky finalized his basic theories of rocket propulsion and control of satellites in space. A few years later, he suggested a method of reaching escape velocity by using a multistage booster, consisting of separate rockets lashed together and launched simultaneously. These very last calculations about multistage boosters led Tsiolkovsky to the conclusion that the first spaceflights would take place within twenty to thirty years. He made this prediction during his last radio speech from Moscow on May 1, 1932. "The Citizen of the Universe," as he called himself, died in 1935.

The Kremlin originally selected the centennial of Tsiolkovsky's birthday—September 17, 1957—to launch Sputnik. Although the successful launch was seventeen days late, the tribute underscored the high esteem that the Russians had for Tsiolkovsky.

The United States paid little or no serious attention to Tsiolkovsky. In fact, the country was equally indifferent to its own firstborn rocket scientist.

*His first sketch of a spaceship showed fuel tanks containing liquid oxygen and liquid hydrogen, the same fuel used in the Saturn V Moon rocket and in the U.S. space shuttle.

Rocket Man

The second pioneer in modern rocketry was Robert Hutchings Goddard. As a sickly child growing up west of Boston, Goddard experimented with kites and telescopes and studied the writings of H. G. Wells and Jules Verne. One day in 1899, while in his teens, he climbed a backyard cherry tree and, as he trimmed the branches, stared upward and daydreamed about a vehicle flying past the Earth's atmosphere to Mars. Years later, Goddard said that when he came down from the tree he was a boy obsessed with the idea of spaceflight, even dreaming of going to the Moon.

Goddard began his study of liquid propellant rockets in 1909. Three years later he proved that rockets would work in a vacuum. He was granted the first two of his many rocket patents in 1914. These patents—and two more issued the same year—covered all the broad principles of rocket engineering, including the feeding of successive amounts of propellant into the combustion chambers to give steady propulsion; the use of multiple rockets, each being dropped as its propellant was exhausted; and tapered exhaust nozzles to take full advantage of the force of the expanding gases.

Armed with these patents, Goddard approached the Navy with the suggestion that rockets could be useful for both airborne artillery and ground-based antiaircraft weapons. The secretary of the Navy replied that he might consider testing some of Goddard's sample rockets.

Next Goddard set about to prove experimentally that a rocket could operate in space. In a large test chamber, built so it could be evacuated of air, he fired more than fifty small rockets. They readily lifted themselves in the vacuum. This was the experimental proof of his theories that he needed. By 1916, he had reached the point where he needed financial help. Two years later, the U.S. Army Signal Corps commissioned him to develop a military rocket and study long-range solid-fuel bombardment rockets. In November of that same year he successfully demonstrated a solid-propellant, recoilless rocket to the Army at Aberdeen Proving Ground in Maryland. This small rocket, launched from a rack normally used to hold sheet music, demonstrated the basic idea of the bazooka. The end of World War I came four days later, ending Army funding for Goddard's rockets, but the bazooka would come into widespread use in World War II.

In May 1919, while studying for his doctorate at Clark University in Worcester, Massachusetts, Goddard sent an article titled "A Method of

Dr. Robert Goddard, the Father of American Rocketry, with a steel combustion chamber and rocket nozzle in 1915. (NASA)

Reaching Extreme Altitudes" to the Smithsonian Institution and soon received a grant of $5,000. This article laid the theoretical foundation for future rocket development in the United States. It also mentioned—almost as an aside—that a rocket could be flown to the Moon as a demonstration and that its arrival could be marked with an explosion that could be seen on Earth with a telescope. The article was published by the Smithsonian in the *Smithsonian Miscellaneous Collection* (vol. 71, no. 2), and nothing happened for some days until a Smithsonian publicist issued a press release devoting a few paragraphs to Goddard's idea of exploding a charge of flash powder on the Moon.

Overnight Goddard became the "Moon Rocket Man." Even dignified newspapers chided him for imitating Jules Verne in writing such fantastic gibberish. "Modern Jules Verne Invents Rocket to Reach Moon" was one headline.

A moment of intense personal dismay came for Goddard on January 13, 1920, when the *New York Times* mounted a tart and ill-informed editorial attack on him and his ideas. It was a follow-up to a previous day's news story titled "Believes Rocket Can Reach Moon." The editorial opined that space travel was impossible for the simple reason that without an atmosphere to push against a rocket, it could not move an inch. Goddard clearly "lacked the knowledge ladled out daily in high schools." The *Times* editorialist was dead wrong, of course, because high schools taught Newton's third law, on which Goddard's premise was built. In 1969, as Apollo 11 lifted off for the Moon, the *Times* alluded to the original editorial in a new one: "Further investigation and experimentation have confirmed the findings of Isaac Newton in the 17th century, and it is now definitely established that a rocket can function in a vacuum as well as in an atmosphere. The *Times* regrets the error." The *Washington Post* termed it "one of the great retractions of our time."*

In the wake of the *Times* attack, Goddard became infamous to the public and press—fascination with this Moon man would not be dampened. The reaction in Europe to Goddard's Smithsonian report was slightly more favorable, but an article in the London paper the *Graphic* insisted that the velocity of Goddard's rocket would cause it to "vanish in an incandescent wisp of flame and smoke."

Soon Goddard's name was tied to all sorts of stunts involving manned spaceflight and trips to the Moon and beyond. Goddard got the following telegram from a publicist at the Mary Pickford Studios in Hollywood:

> WOULD BE GRATEFUL FOR OPPORTUNITY TO SEND MESSAGE TO MOON FROM MARY PICKFORD ON YOUR TORPEDO ROCKET WHEN IT STARTS.

More than 100 people volunteered for the flight. An aviator from Philadelphia said he would even let Goddard send him to Mars if he would provide him with $10,000 in life insurance. A *Baltimore Sun* article claimed that Ruth Phillips, a Kansas City woman, was willing to outdo the suffragettes by joining the flight to Mars, demonstrating the ultimate in women's independence. The *New York World* carried the story of Captain Charles N. Fitzgerald, a daredevil pilot who wanted to be "the first visitor to say 'Hello' to the man in the moon."

*Some have blamed the *Times* attack for the way Goddard lived the rest of his life. In 1999, in a special edition of *Time* devoted to the greatest minds of the century, Jeffrey Kluger said that after the attack, Goddard "sank into a quarter-century sulk from which he never emerged."

Unable to rebut any of this nonsense, the self-effacing, proud New Englander quickly learned to shun publicity and grant few interviews because whatever he said or did was turned into a lurid headline. For example, in April 1920 he talked to the Chicago chapter of the American Association of Engineers about the future possibilities of rockets and mentioned that radiation in space might be a problem; the *Milwaukee Journal* slugged its story "Moon Beams Would Cremate Human Rockets!" As late as 1945, weeks before his death, Goddard wrote, "The subject of projection from the earth, and especially a mention of the moon, must still be avoided in dignified scientific and engineering circles."

Working in secret, he developed ideas and rockets that grew in complexity. While working on solid-propellant rockets, Goddard became convinced that a rocket could be propelled better by liquid fuel. No one had ever built a successful liquid-propellant rocket before, because it was a much more difficult task than building a solid-propellant rocket. Fuel and oxygen tanks, turbines, and combustion chambers were needed. In spite of these difficulties, on March 16, 1926, he stood in a snow-covered field owned by Effie Ward, a distant relative, on a farm in Auburn, Massachusetts, and, with the aid of a blowtorch, launched the world's first liquid-fuel-propelled rocket. The ten-foot-long projectile soared to forty-one feet at sixty miles per hour, landing in a frozen cabbage patch 184 feet away. For several years he did not publicize this defining moment in the early history of spaceflight lest he be mocked by newspaper editorialists.

As the rockets got bigger, the farm became a less appropriate launchpad. His neighbors were afraid that he would do them harm. A July 17, 1929, launch was so loud that it brought out a police car, two ambulances, and a search plane. The fire marshal for the commonwealth looked into banning Goddard from further rocket testing in Massachusetts.

Goddard was finally cleared for an isolated area of the Army's Fort Devens. This was, for Goddard and his crew, a successful—though noisy—launch. Scientists had long known that in order to learn more about the earth they had to take their instruments to higher altitudes to reduce the absorbing effects of the atmosphere. Goddard's payload—a barometer, a thermometer, and a camera—was sent to an altitude of ninety feet and then parachuted back to Earth.*

*To his neighbors and the press, success equaled failure, and anything short of a shot into the stratosphere was no more than an "explosion"—even this milestone. One Worcester paper led with "Terrific Explosion as Prof. Goddard of Clark Shoots His Moon Rocket,"

At the end of the 1920s, Goddard still did not have adequate financing for his experiments. Fortunately, Charles A. Lindbergh helped him get a series of grants from the great benefactor of aviation Daniel Guggenheim and later the Daniel Guggenheim Fund for the Promotion of Aeronautics. With this money Goddard was able to buy a large tract of land in Eden Valley near Roswell, New Mexico, where he could work in seclusion.

In the 1930s he was considered the world's top rocketeer. He amassed 212 patents, all dealing with aspects of the exploration of outer space, including liquid-fuel combustion, engine-cooling systems, and gyroscopic steering. Goddard, like Tsiolkovsky, believed that multistage rockets, or steprockets, were the answer to achieving high altitudes and that the velocity needed to escape Earth's gravity could be achieved in this way. Because he rarely published the results of his work and avoided publicity, his influence was limited. Hungarian-born scientist Dr. Theodore von Kármán, director of Caltech's Guggenheim Aeronautical Laboratory, tried to persuade Goddard to join his new Rocket Research Project, a band of rocket researchers, but Goddard declined the invitation.*

In March 1937 a Goddard rocket rose to between 8,000 and 9,000 feet from the New Mexico desert. G. Edward Pendray, one of the founders of the American Rocket Society, later wrote that this rocket "contained almost all of the features (. . .) incorporated into the German V-2 rockets—though the latter were much larger."

When Goddard saw that war was imminent, he offered himself to the military. In May 1940, Harry Guggenheim, Daniel's son and the head of the Daniel and Florence Guggenheim Foundation, arranged a meeting in Washington, D.C., between Goddard and representatives of the Army, Navy, and Air Corps to present the military possibilities of long-range liquid-propellant rockets. Guggenheim was told that the military saw no possibilities for God-

---

while another smugly proclaimed, "Moon Rocket Misses Target by 238,799½ Miles." The *New York Times* worked up a double-decker headline: "Meteor-Like Rocket Startles Worcester," followed by "Clark Professor's Test of New Propellent to Explore Air Strata Brings Police to Scene."

*The original goal of the von Kármán group was to develop a small rocket that would be a means of studying the atmosphere, but as World War II got under way, they turned to developing rocket motors to assist heavy warplanes in takeoff. The Caltech group reorganized in 1944 as the Jet Propulsion Laboratory, where a von Kármán–led team went on to become the first group in the United States to build and launch a rocket specifically designed for upper-air research. Named the WAC-Corporal, the rocket became a significant launch vehicle in postwar rocket research.

Robert Goddard and company with the rocket used in the New Mexico test of September 23, 1931. Left to right are: N. Ljungquist, L. Mansur, Goddard, A. Kisk, and C. Mansur. (NASA)

dard's guided missile and was informed by an Army representative that the "next war will be won with the trench mortar." During the war Goddard worked for the Navy in Annapolis, Maryland, on JATO (Jet-Assisted Take-Off) rockets to help shorten takeoff lengths for heavily loaded seaplanes. It was, as Pendray later wrote, research for a "pedestrian, unimaginative and largely futile purpose." He added, "It was like trying to harness Pegasus to a plow."

As late as the bombing of Pearl Harbor in 1941, the United States did not have a single rocket weapon in its arsenal, although the country was getting into rocketry slowly and cautiously. In 1943, the Army established the Ordnance Rocket Branch to manage the development of all rockets for the United States. "Goddard," wrote his biographer Milton Lehman, "was not invited to join." A year later, the Ordnance Rocket Branch signed a contract with the California Institute of Technology's Jet Propulsion Laboratory to study rocket propulsion and develop long-range surface-to-surface rockets.

This became known as Project ORDCIT, which developed and tested twenty-four solid-propellant rockets at Fort Irwin, California. The U.S. Army developed the Private, Corporal, and, later, Bumper rockets. These were developmental systems that never reached the stage of operational testing during the war.

Goddard, a man who loved to smoke his cigars down to the nub, died on August 10, 1945, of throat cancer and was laid to rest in the Goddard family plot in Worcester on the fourteenth, the day the Japanese surrendered, ending World War II. Before he passed away, he told a reporter from the *Worcester Telegram*: "I feel we are going to enter a new era. . . . It is just a matter of imagination how far we can go with rockets. I think it is fair to say you haven't seen anything yet." There was irony, perhaps unintended, in this statement because all during World War II the United States totally ignored Goddard and his work with liquid rockets and flight control. Germany, on the other hand, had paid much more attention to rockets in general and to what Goddard was doing.

Some modern historians have called Goddard secretive, thin-skinned, too sensitive; but such characterizations are unfair and miss the point. Goddard was a tough, resilient man working under dangerous conditions with little compensation. He needed room and a place to experiment. As A. Scott Berg wrote in *Lindbergh*: "Worse for Goddard than being dismissed as a lunatic was the undue attention being placed on his experimentation, which stripped him of the privacy necessary for trial and error." This was precisely why Lindbergh and Guggenheim helped him move west. As for his covert behavior, his wife, Esther, put it in perspective: "Robert isn't secretive. He just doesn't like to talk too much." He did report his findings to those who supported him—namely, Clark University, the Smithsonian, the Guggenheim family, and Lindbergh.

In the days following the 1957 Sputnik launch, Esther Goddard reminded Americans that in the 1920s, "many foreign nations, including Russia, Japan, Germany, and Italy, wrote to my husband asking for his services, but he turned them all down even though he received little support from his own government after World War I."

Goddard was increasingly forgotten in his own country in the years following his death. America was eventually forced to recognize its native rocket genius through a lawsuit. The Guggenheim Fund, which had been a prime means of support for Goddard and with whom he had shared ownership of his patents, sued the federal government for patent infringement in 1951. In 1960, fifteen years after his death, the U.S. government acknowledged Goddard's work when the Department of Defense and the National Aeronautics and Space Administration awarded Esther Goddard and the foundation one million dollars for the use of his patents, which had been applied in designing the Atlas, Thor, Jupiter, Redstone, and Vanguard rocket engines—the power that got America into space. Money recovered by the Guggenheim group was used to fund Goddard professorships at Princeton and Caltech.

History eventually caught up with Robert Goddard. Harry Guggenheim said of him after the first manned Moon landing, "He was to the moon rocket what the Wright brothers were to the airplane." Biographer Lehman summarized him as follows: "Robert H. Goddard—sickly, frail, unrecognized and even derided—accomplished more in a lifetime working almost alone than, perhaps, any other inventor of the twentieth century. He, more than any man, opened the door to the Space Age."

On May 1, 1959, the NASA facility in Beltsville, Maryland, was named the Goddard Space Flight Center in his honor. He was awarded a posthumous congressional gold medal in the same year, and a postage stamp was issued in his honor. The stamp prompted his biographer, Milton Lehman, to write an article in the *Washington Post* titled "So We Issued a Stamp," which carried the subheading "Robert Goddard Invented Rocketry 50 Years Ago But Only Germans and Russians Paid Heed." His works were studied abroad, especially in Russia and Germany. For many years his basic rocket patents could be bought in a bundle from the patent office for a mere fifteen cents. But in the final analysis, it was a combination of Goddard's need for secrecy and his nation's disinterest that allowed others to build on and profit from his work. The branch of rocketry that would someday land Americans on the Moon would stem from a different tree—albeit one that developed in the same direction.

The Transylvanian Motivator

The last of the rocket pioneers was Hermann Julius Oberth, the only one of the three to live to see his dreams realized. At the age of seventy-five, he was an honored guest at the July 1969 launch of Apollo 11, and at ninety-one, at the October 1985 launch of the space shuttle. Oberth,

Rocket pioneer Hermann Oberth. (NASA)

a gangly mathematician from Transylvania in Romania, grew up reading Verne and Wells and tinkering with models. When he was fifteen, he built a small replica of the rocket engine described in Verne's *From the Earth to the Moon.* He became a very public man, the opposite of the isolated Tsiolkovsky and the publicity-shy Goddard. Twenty years after the Tsiolkovsky equation was first published, Oberth had worked out similar formulas on his own in Germany.

Ironically, Goddard's unwanted publicity may have had the effect of energizing rocket science in Germany and other countries at a time when it appeared all but dead. On May 3, 1922, while a student in Heidelberg, Oberth wrote to Goddard requesting a copy of his treatise, which Oberth had read about in a newspaper article on Goddard. The letter made Goddard uneasy because, according to Milton Lehman, he was suspicious of the German "tendency to turn inventions into weapons of war," but Goddard sent the copy nonetheless.

The next year, thirty-year-old Oberth published *The Rocket into Planetary Space,* which contained the bold thesis that it was then possible to build liquid-propellant rockets capable of escaping the Earth's atmosphere. He said that within a few decades manned space stations and interplanetary flight would also be possible. Goddard found Oberth's treatise to be disturbing and would feel this way for many years. He knew that Oberth was brilliant but could never get over his belief that the younger man had borrowed heavily from him.

Oberth's book meanwhile helped inspire an international astronautics movement in the 1920s in Germany and elsewhere. In 1925, for instance, Walter Hohmann, a German, published his book *The Attainability of Celestial Bodies,* in which he set forth the principles of rocket travel in space, including how to change the orbit of satellites. The "Hohmann transfer" is now a routine procedure to get payloads from low Earth orbit into a geosynchronous orbit—one that moves at the same rate as the Earth turns.

In 1928, Oberth completed his doctoral dissertation, "The Road to Human Space Travel." Though rejected as an academic thesis, this 423-page document was circulated widely throughout Europe and especially in Germany. Oberth suggested that if a rocket could develop enough thrust, it could place a payload into orbit. He pioneered concepts as diverse as docking spaceships and sending a telescope into space for improved viewing.

A tireless promoter of space travel, in 1928 he was retained by the great filmmaker Fritz Lang as adviser to a space film called *Frau im Mond* (Woman on the Moon), in which he also appeared. Oberth was also commissioned to create a rocket that would be launched at the movie's 1929 premiere. The attempt never got off the ground, mainly because of Oberth's shortcomings as a practical engineer, but it helped spur interest in rocketry in Germany.*

Because of the movie and his new book, Oberth soon had a band of followers, including an eighteen-year-old named Wernher von Braun, the second of three sons born to Baron Magnus von Braun and Baroness Emmy von Quistorp. Oberth's protégé, von Braun would become the most important rocket developer and champion of space exploration of the twentieth century.

*The film's script included the now-famous reverse countdown—5-4-3-2-1—before ignition and liftoff.

CHAPTER THREE

Vengeance Rocket

I kept my eyes glued to the binoculars . . . it was an unforgettable sight. In the full glare of the sunlight the rocket rose higher and higher. The flame darting from the stern was almost as long as the rocket itself. . . . The air was filled with a sound like rolling thunder.

—Nazi rocket boss Walter Robert Dornberger describing the October 3, 1942, first launch of the 13.5-ton A-4 rocket, later called the V-2 rocket

Rocket clubs sprang up throughout Germany in the late 1920s as fascination for outer space grew. The leading club was founded in June 1927 and called the Verein für Raumschiffahrt (Society for Space Travel). Its president and main source of inspiration was Hermann Oberth, who had by this point developed a wide following in Germany and abroad. Membership in the VfR rose steadily, and it was soon publishing books and demonstrating a motley assortment of rocket-powered railroad cars, sleds, and automobiles.

These rocket demonstrations attracted worldwide media attention. For instance, the Opel-Rak 2, an automobile outfitted with twenty-four rockets and short, inverted wings positioned to keep the car from leaving the ground, reached the speed of nearly 125 miles per hour on May 23, 1928. At the wheel of the vehicle was automobile tycoon Fritz von Opel, who wanted publicity for his conventionally powered passenger car.

Oberth's ability to publicize rocketry through photogenic and newsworthy experiments transferred to others, including Goddard, who was no longer perceived as such an oddity. Milton Fairman wrote in the March 1930 issue of *Popular Mechanics*: "Thus the plans of Prof. Robert H. Goddard, of Clark University, and the German scientist Hermann Oberth are being received with higher regard than were the first pronouncements of Doctor Goddard some twenty years ago. The scientific world is becoming rocket-conscious."

And so was the German military. In the late 1920s and early 1930s, the German generals were trying to rearm their country with a weapon that would not violate the Versailles Treaty, which ended the First World War. The drafters of the treaty had worked hard to keep Germany disarmed but made a historic blunder by omitting rockets from their long itemized list of weapons Germany could not develop. The omission would change the course of history.

Germany's Formidable Rocket Program

German artillery captain Walter Robert Dornberger was assigned the task of looking into the feasibility of using rockets as weapons. He visited Oberth's Society for Space Travel and was so impressed with the group's enthusiasm and dedication that he gave them a contract and a fee of $400 to build a single rocket. Wernher von Braun, who had joined the VfR in September 1929, worked through the spring and summer of 1932, only to have that rocket fail when tested in front of military observers. However, Dornberger was so taken with von Braun that he hired him as a member of the military's new rocket artillery unit, located in Kummersdorf, about sixty miles south of Berlin.

Von Braun's natural talent as an engineer and his boundless exuberance were apparent to all, and, despite his young age, he was soon leading the research team for Dornberger. The first new rocket developed at Kummersdorf was the A-1, an abbreviation for Aggregate 1. The first A-1 blew up. The design was reconfigured as the A-2. In December 1932, two A-2 rockets, named Max and Moritz after comic characters in a German book, made back-to-back flights to altitudes of approximately 6,500 feet. The success of the A-2 got the Dornberger–von Braun team more funding to hire more engineers.

The following year, Adolf Hitler came to power, with Hermann Goering

Germany's rocket pioneers in 1930 with a rocket designed by Hermann Oberth (fifth from right). Wernher von Braun is the second man from the right. (NASA)

heading his air force, the Luftwaffe. Both men saw great potential in rockets as pilotless weapons of large-scale destruction, though not for a moment as spacecraft. By the beginning of 1934, they had a team of eighty engineers building rockets in Kummersdorf. At this point Oberth's pioneering days were over, and his beloved VfR was quickly failing because of unpaid bills. More to the point, the VfR had incurred the chilling displeasure of the Gestapo over its ties to rocketeers in other countries. However, Oberth's influence continued to affect the work of von Braun and his powerful team of rocket builders.

Kummersdorf soon proved too small for the growing team and much too close to populated areas for their increasingly dangerous war rockets, so a new facility had to be built. Usedom, an isolated area on the coast of the Baltic Sea and due south of Sweden, not far from what is today the border of Poland, was chosen as the new site. The military complex was named Peenemünde after a tiny fishing village nearby. It was large enough for the launching and monitoring of rockets over ranges up to about 200 miles.

By the spring of 1936, work was under way on the world's first fully self-contained research-and-development center with everything needed to design, build, and launch large rockets. When complete, it housed two distinct and competing programs, in separate wings. Peenemünde West was a relatively small Luftwaffe facility where the Fieseler-103 flying bomb was tested, while Peenemünde East was home to von Braun's program.*

The Rocket Team, as von Braun's group was called, developed A-3 rockets but met with initial failure. Four test models of the new twenty-one-foot-long, 1,650-pound A-3 failed, and, according to munitions minister Albert Speer, Hitler was "having grave doubts" about rockets as weapons. Von Braun and company moved on to the A-5, which in October 1938 reached an altitude of eight miles. But as the Nazis invaded neighboring countries, beginning with the annexation of Austria, altitude became less important than the ground-to-ground rockets being developed by the von Braun team and the Luftwaffe to be used to deliver warheads loaded with high explosives.

Funding continued to flow to von Braun's team, which launched some seventy A-3 and A-5 rockets between 1937 and 1941. They were testing components for use in the proposed A-4 rocket, which would soon be known to the world as the V-2, a single-stage rocket fueled by alcohol and liquid oxygen. The V-2—the *V* stands for *Vergeltungswaffen*, or "revenge weapons," the popular name for the A-4—stood 46.1 feet high and had a thrust of 56,000 pounds. The first one flew in March 1942 before crashing into the sea a half mile from the launch site. A second launch in August rose to an altitude of seven miles before exploding. The third try was on October 3 when the first V-2 successfully test-launched from Peenemünde. Breaking the sound barrier, it reached an altitude of sixty miles, followed its programmed trajectory perfectly, and landed on target. It was, at once, the world's first launch of a ballistic missile, the first rocket ever to go into the fringes of space, and the ancestor of practically every rocket flown in the world today. It was a true spaceship in that it carried both its own fuel and oxygen and could, if needed, work in a vacuum. But it was also a formidable weapon capable of carrying a warhead of 2,200 pounds; the silent V-2 could reach a velocity of 3,500 miles per hour, making it almost impossible to shoot down.

While the V-2 was being developed, the Luftwaffe worked on the Fieseler-103. It was first successfully tested in 1941 and was renamed the V-1. Leaving nothing to chance in his campaign to punish those who had

*Initially, Oberth was not allowed to go to Peenemünde as he did not hold German citizenship and had become tangled in Nazi suspicions and the military's inter-service rivalry. The Romanian professor was, as he termed it, "kept on ice"—occupied with trivial tasks for the Luftwaffe—and was not transferred to the Baltic rocket base until the fall of 1942, after the V-2 was already completely tested and ready for production.

opposed him in England, Hitler ordered 25,000 V-1s. Carrying over a ton of high explosives, the V-1 moved just under the speed of sound and was little more than a pilotless plane that flew at 350 to 400 miles per hour for 150 miles. By June 1944, enough V-1s had been produced to start the attack on southern England. More than 8,000 "buzz bombs," as the British called them because of their telltale noise, were launched against London alone. Thousands more were fired against Allied targets on the Continent. Even though the V-1 carried about the same amount of explosives and had about the same range, it was an inferior and unreliable weapon compared to the silent, supersonic V-2 being built in converted Volkswagen factories. For example, only 211 of the 5,000 V-1s fired against Antwerp ever exploded on target.

In 1943 Hitler told von Braun and Dornberger that the Third Reich required a super "weapon of vengeance." Von Braun showed him a movie of the V-2 in operation and said that it could carry more than a ton of TNT across the English Channel into London and its suburbs. Hitler decided to use the V-2 as his superweapon and ordered 12,000 of them.

On March 15, 1944, von Braun was arrested by the Gestapo and imprisoned in Stettin prison. According to Ernst Stuhlinger and Frederic I. Ordway III in their von Braun memoir: "His alleged crime was declaring that his main interest in developing the A-4 was its potential for space travel rather than as a weapon that would turn the tide of war in Germany's favor. Also, since he regularly piloted a government-provided airplane, it was suggested he might be planning to escape with A-4 secrets to the Allies." It was only through the intervention of munitions and armament minister Albert Speer and Major General Walter Robert Dornberger that von Braun was allowed to return to work, because the success of the V-2 program depended on him.

The Rocket Team was now using slave labor, working in underground factories in the Harz Mountains, to mass-produce V-2s to rain explosives on London. On September 7, 1944, the first combat V-2 was launched and landed in Chiswick, London. Some 4,300 V-2s were eventually launched against England, Belgium, and other Allied targets. Once a V-2 was launched, the Allies had no way to intercept it since it fell silently on its target at about 5,200 feet per second, about the same speed as a modern tank round. When the first of them hit London, they were believed to be exploding gas mains. Fortunately for the Allied forces, the V-1 and V-2 came too late in the war to save the Third Reich, but the two V weapons together killed 8,938, seriously wounded another 25,000, and damaged more than one million homes.

America's Quest for Remote-Controlled Bombs

On a much less ambitious level, the United States had also been looking for airborne weapons that did not require pilots. From 1928 to 1932, the American effort focused on turning conventional aircraft into working, unmanned drones for gunnery practice. Then, in late 1932, there was a sudden rush of military interest in remote-controlled vehicles, which led to a small menagerie of "special weapons"—including the "Bug," essentially a surface-to-surface "buzz" bomb, and the "Bat," a radio-controlled glide bomb. Out of this pack came the crude but lethal GB-1, which was a 2,000-pound bomb with plywood wings, rudders, and a radio-control package. GB-1s were dropped from lumbering B-17 bombers and guided to their targets by bombardiers using remote control.* Later in the war came the GB-4 "Robin," the first guided weapon to use a television camera as its "eyes." Americans were learning to control objects in midair—in other words, to guide missiles.

Germany's V-2s were a thorn in the side of the Allies, and a new strategy was required to deal with them. As the air war progressed, B-17 and B-24 bombers literally began to wear out. These gigantic surplus bombers occupied valuable hangar space and even more valuable maintenance time. By late 1943, General H. H. "Hap" Arnold directed Brigadier General Grandison Gardner and his engineers at Eglin Field in Florida to strip these hulks, nicknamed Weary Willies, of all but their most essential equipment to make room for high explosives. Human crews were to get the planes in the air, then bail out, and a manned mother ship would take remote control.

One of the most tragic moments in the early days of airborne robots came late in the war. A top-secret American effort—known as Project Aphrodite and based in England—planned to use Weary Willies to destroy the German rocket research and launching sites concealed in caves and heavily fortified bunkers in and around Peenemünde. The highly explosive bomber could presumably zero in on the mouths of the caves and large

*In 1943, 108 GB-1s were dropped on Cologne, causing heavy damage. For propaganda reasons—perhaps mixed with a genuine inability to comprehend what had happened—the Luftwaffe officially reported that it was able to turn back the bombers before any bombs were dropped. The report continued: "The accompanying fighter cover, however, composed of small, exceedingly fast twin-tailed aircraft, came over the city in a strafing attack. So good were the defenses that every single fighter was shot down; much damage was done by these falling aircraft, all of which exploded violently."

bunkers guided by the combination of TV eyes in the tip of the robot and the radio remote-control system.

More V-1s and V-2s were being fired at London after D-Day, June 6, 1944, and the urgency of the project was further intensified by dark suspicions on the part of American intelligence that the Nazis were preparing a new series of V rockets for an attack on New York City. The A-9 and A-10 rockets never got beyond the planning stages, but the Allies did not know this.*

Project Aphrodite was a complete failure. A number of planes did not even get near their targets, and those that did somehow managed to miss their mark at the last moment. The financial cost was high, and a number of pilots and crewmen died during the effort, including young Joseph Kennedy Jr. and a crewman when their B-17 exploded before they were able to bail out. The explosion was so great that fifty-nine buildings in the British small town of Newdelight Wood were damaged. Colonel Elliott Roosevelt, the president's son, was following in a small photo-reconnaissance plane to take pictures of the operation, and the explosion came close to destroying his plane as well.

The details of Joseph Kennedy's death were officially kept top secret for years, and it was not until his brother John became president that the full circumstances became known to the family. President John F. Kennedy made plans to meet with the Texas family of the young man who had died with his brother. The meeting was to have been on November 22, 1963, the day he was killed in Dallas.

Ironically, if Joseph Kennedy had been able to create major damage at Peenemünde, he might have killed members of the Rocket Team, who were later to be central to John F. Kennedy's plan to reach the Moon. In fact, one member of that team, Walter Thiel, who was in charge of V-2 engine development, had been killed the year before Aphrodite, when hundreds of British heavy bombers had attacked Peenemünde, killing 800.

The importance of Project Aphrodite was not its impact on the outcome of the war, for General Arnold had no great hopes for the ultimate decisiveness of these weapons; rather, it furthered the notion that planes without pilots would be a key factor in the future of warfare. As the war wound down, it was clear to the American military that the concepts at work in Aphrodite,

*The design of the A-9 called for supplementary wings that would enable it to glide when the propellant had been exhausted, attaining a range of some 300 miles. The A-10 was to be a giant eighty-seven-ton booster rocket that would carry the A-9 above the atmosphere, at which point the A-9 would proceed under its own power for a distance of some 2,500 miles.

the gasoline-burning V-1, and the V-2 ballistic rocket would lead to the immediate development of guided missiles.

The United States wanted the German V-2 rocket, the most sophisticated and capable rocket that had ever been created. Once launched, it was a formidable weapon that could devastate whole city blocks. American forces in and around London and those who had come ashore in the D-Day invasion of Normandy had firsthand experience with the V-2s. One of them was John B. Medaris, a spit-and-polish Army officer who went overseas in the spring of 1942 and served as an ordnance officer in the campaign in Tunisia and the invasion and occupation of Sicily. He transferred to England, where he was part of the team that planned and executed the D-Day invasion. In France, he remained with the First Army throughout the remainder of the campaign in Europe. One day in the early autumn of 1944, a stray V-2 fell short of its intended target and exploded close enough to Medaris's headquarters to rattle windowpanes and shake down dust from the rafters. The rocket landed in a ravine near the headquarters' cook tent, and because the ravine absorbed the blast there were no fatalities, although several cooks were injured.

Having seen the destructive power of the V-2 and knowing that America did not have its counterpart—or anything resembling it—on its drawing boards, the Army planned to get hold of as many V-2s as it could, along with its top engineers. The United States correctly assumed that the Russians had much the same idea; the scramble for the human and hardware spoils of the V-2 would become one of the first contests of the Cold War.

The German Rocket Team Surrenders

In late January 1945, Hitler's Third Reich was collapsing. As the Russian army approached and its guns could be heard, von Braun immediately assembled his planning staff and asked them to decide how and to whom they should surrender. Most of the Rocket Team was frightened of the Russians and reasoned that war-ravaged Britain did not have enough money to afford a rocket program and would not be especially keen on one led by V-2 engineers. They felt the French might treat them like slaves in retribution for using members of the French Resistance as slave laborers to build the V-2. The thought also occurred to them that there might be an attempt to try them as war criminals. That left the Americans, who the Germans felt needed them, could afford them, and would rather let them build rockets than put them on trial. The Rocket Team knew they had to make a deal with the Americans as quickly as possible.

The team had reason to fear the wrath of the Allies, for no amount of space talk could expunge the lethal record of the V-2s. In the final days of the war, V-2s were launched in a stepped-up terror campaign against civilians in Allied cities. Also, there had been many heavy casualties during the manufacturing of the rockets in Mittelwerk, a vast underground factory in central Germany where hazardous work was done mainly by slaves drafted from nearby concentration camps.

"The V-2 was unique in the history of modern warfare, in that more people died building the rockets than lost their lives on the receiving end," wrote Tom D. Crouch, senior curator of aeronautics at the Smithsonian's National Air and Space Museum, in his book *Aiming for the Stars*. "Between 1943 and 1945, 60,000 slave laborers passed through the gates of Dora and other concentration camps associated with Mittelwerk. One-third of them—20,000 human beings—died there. There is no accounting of the additional thousands of slave laborers who perished at the liquid-oxygen plants and other facilities related to V-2 operations."

Besides wanting to work with the Americans in space, von Braun—although himself a major in the Schutzstaffel (SS)—wanted to move south quickly because he feared that his team might be massacred by the SS to keep Nazi secrets from being passed to the Allies. He also feared the team might be used by the SS as chips to bargain with the Allies. The Rocket Team had been heavily guarded for years. At the height of the Peenemünde era, more than 4,000 men and women were stationed there, but, according to British historian Jeffrey Robinson, more than half were SS guards.

Von Braun directed an incredible exodus through a failing Germany to surrender to the Americans. By faking orders and bluffing, he got three trains and 100 cars and trucks loaded with equipment, engineering drawings, and research papers. From the time the first train left on February 17 until their surrender, the group faced many crises, including a car crash that left von Braun with a broken arm. They hid the most important documents in a mine shaft and evaded the dreaded SS while searching for the Americans. The elite German Rocket Team was holed up in a ski lodge in Bavaria when it got the word that Adolf Hitler was dead and the Third Reich was crumbling. In the final days of the war, the Allies had put a great premium on destroying the operation at Peenemünde, along with the scientists responsible for the reign of destruction.

With elements of the advancing U.S. Seventh Army only a few miles away on May 2, 1945, Wernher von Braun's younger brother Magnus climbed on a bicycle and set off down a country road in search of the Americans. He was picked for this delicate mission of surrender because he spoke the best English. He found Private Fred Schneikert of Sheboygan, Wisconsin, who pointed an M-1 rifle at him. The twenty-five-year old German said that the inventor of the V-2 was nearby, ready to surrender. Schneikert later recalled him saying, "I think you're nuts"; however, he relayed the information. Arrangements were made for the von Braun party to come through the lines that night.

In one of the most dramatic struggles of the war's final days, the Americans raced the Soviet army for the spoils of the V-2 program. Realizing the importance of these engineers and noting that the Russians had in their sights Peenemünde and other strategic rocket locations, the Americans immediately went to an underground storage site at Nordhausen, in what was soon to be the Soviet zone of occupation, and captured enough parts to fabricate 100 V-2 rockets. The parts filled 360 commandeered railroad cars, which were transferred to the hulls of sixteen Liberty ships and sent directly to the United States.

Three days after the Americans took Nordhausen, Peenemünde fell into the hands of advancing Soviet forces, who captured a considerable amount of hardware and production facilities, but almost all of the important German technical personnel and documents were already in American hands. Much of von Braun's production team was captured by the Russians, but the Americans had gotten the brainpower, much of the hardware, and almost all of the remaining V-2s and spare parts. The Russians contracted with those V-2 technicians it could obtain and let it be known that there would be a bonus of 50,000 marks if Wernher von Braun would come and work for the Communists.

The Rocket Team Crosses

Members of the Rocket Team and their families were guests of the Army and were housed in a compound near Garmish-Partenkirchen that was built to house athletes at the 1936 Winter Olympics. In those early days of captivity the Germans were being culled to find the most talented who were prepared to be brought to the United States. Out of the more than 1,000 who had deserted Peenemünde, just over 100 would be offered the rare opportunity of transferring directly from one army to another.

U.S. forces obtained the fourteen tons of blueprints and plans that the Rocket Team had hidden in the mine shaft. These included 3,500 reports and 510,000 engineering drawings for the V-2 and more advanced rockets. According to Tom D. Crouch, "Allied interrogators, each one anxious to uncover the secrets of the Nazi-weapon program, flocked to the area." Unknown to the Rocket Team until after their surrender was the fact that the U.S. Army was looking for them to make a human transfer of technology as part of "Operation Paperclip," so called because the paperwork of those picked to come to the United States by Army Ordnance was marked by paper clips.*

From the first day of his surrender, Wernher von Braun appeared to be in full command of his team as well as his captors. According to one American account, he "treated our soldiers with the affable condescension of a visiting Congressman . . . posing for endless pictures . . . like a celebrity rather than a prisoner." Private Fred Schneikert, who had captured the von Braun brothers, was soon trading good-natured gibes with Wernher, who told him he would be in America in a matter of weeks. "I told him I'd be back home before he ever saw America. But I was wrong," Schneikert later recalled.

On June 20, 1945, U.S. secretary of state Cordell Hull approved the transfer of von Braun and a team of more than 100 technicians. U.S. Immigration agreed to look the other way, and on October 1, 1945, the War Department issued a press release that announced: "The Secretary of War has approved a project whereby certain understanding German scientists and technicians are being brought to this country to ensure that we take full advantage of those significant developments which are deemed vital to our national security." The press release went on to describe how the scientists would be interrogated and how the United States would benefit from their knowledge. It ended with the following statement: "Throughout their temporary stay in the United States these German scientists and technical experts will be under the supervision of the War Department but will be utilized for appropriate military projects of the Army and Navy." They would not have access to classified material, and once their job was done—or as von Braun put it, after they'd been "squeezed like lemons"—they would be sent home to Germany.

*On July 19, 1945, the Joint Chiefs of Staff established "Project Overcast," a plan to "temporarily" exploit German scientists. When the code name was somehow compromised, they changed it to "Operation Paperclip." The original idea was to use the Germans in the war against Japan. After Japan surrendered, the project was reexamined and the Pentagon decided it could afford to be more selective and have its pick of the Nazi brain trust.

The German Rocket Team reassembled in the United States at New Castle Army Air Base, just south of Wilmington, Delaware, and after an interim stop in Boston most of them were taken to Wright Field in Dayton, Ohio, for extensive questioning in the summer of 1945. A summary report on von Braun's interrogation underscored the extent to which von Braun and company had been headed toward space. The plan had been to fire several instrumented V-2s straight up into "the top layer of the atmosphere" in the spring of 1945.

The interrogations also underscored how far the United States had allowed itself to fall behind in a new technology that, it had long been assumed but never proved, was aided by Goddard's work. Ironically, Dornberger and others had used news of Goddard's work to suggest that Germany was in a race with America. In a 1963 letter to biographer Milton Lehman, von Braun revealed how useful Goddard had been to the cause: "Goddard's name was known in German rocketry circles in those pre-Hitler days. In speeches aimed at popular support for spaceflight he was frequently cited as an example of how farsighted people were in other countries; in particular, in America."

Goddard himself saw the V-2 as something that should have been American and could have been if he had been given the green light. On a photograph of his last 1941 rocket, Goddard wrote in 1944, "It is practically identical with the German V-2 rocket." Goddard himself got to examine a V-2 before he died, and when his longtime machinist exclaimed, "It looks like ours," he replied, "Yes, Mr. Sachs, it seems so." Fellow rocketeer Pendray reported that one of the German engineers being interrogated about the V-2 in 1945 had said, "You have a man in your country who knows all about rockets, and from whom we got our ideas—Robert H. Goddard."

Von Braun commented in 1959 that Goddard was "ahead of us all," and that "in light of what has happened since his untimely death, we can only wonder what might have been if America realized earlier the implications of his work. I have not the slightest doubt that the United States today would enjoy unchallenged leadership in space exploration had adequate recognition been accorded to him."

From Dayton and other staging areas, von Braun and the other 126 "Peenemünders," as they came to be called, were sent to their new home at Fort Bliss, Texas—a large Army installation just north of El Paso, under the command of Major James P. Hamill. They found themselves in a strange situation as they began their new lives in the United States. Because they could not leave Fort Bliss without a military escort, they sometimes referred to themselves as "POPs," Prisoners of Peace.

Their first months in the United States made for an odd moment in American cultural history. As Constance McLaughlin Green and Milton Lomask wrote in *Vanguard: A History*, this was a time of extreme awkwardness because "the haughtiness of the Germans who landed at Wright Field in the autumn of 1945 was not endearing to the Americans who had to work with them. The Navy wanted none of them, whatever their skills. During a searching interrogation before the group left Germany a former German general had remarked testily that had Hitler not been so pig-headed the Nazi team might now be giving orders to American engineers; to which the American scientist conducting the questioning growled in reply that Americans would never have permitted a Hitler to rise to power."

As the horrors of the slave labor camps were reported, there was some stated resistance to the Germans. On December 30, 1946, a group of eminent individuals, including Albert Einstein, union organizer Philip Murray, and religious leader Norman Vincent Peale, sent a statement of protest to President Harry S Truman. They wrote: "We hold these individuals to be potentially dangerous carriers of racial and religious hatred. Their former eminence as Nazi party members and supporters raises the issue of their fitness to become American citizens and hold key positions in American industrial, scientific, and educational institutions."

One of the very few individual objections is preserved as an undated clipping in the von Braun papers at the Library of Congress. Its author, Rabbi Stephen Wise, who also signed the petition, objected to allowing people whom he regarded as "ardent pro-Nazis" into the country while concentration-camp survivors were refused visas. "Red tape, lack of shipping facilities, and every other possible handicap face these oppressed people," Wise commented in a 1947 letter to the secretary of state, "while their oppressors are brought to this country and are favorably housed and supported at our expense. As long as we reward former servants of Hitler while leaving his victims in D.P. camps, we cannot pretend that we are making any real effort to achieve the aims we fought for."

The extent to which a major in the SS got special treatment extended much farther than housing and wages. At the end of the war von Braun's parents were snatched out of the Russian-controlled Poland by covert means. Deeply moved by this remarkable act, Wernher and his brother Magnus wrote these words to the U.S. Army officer, James P. Hamill, who pulled off the rescue and who was to become the man in charge of the Rocket Team at Fort Bliss: "Having received the message that our parents entered Landshut, we want to express thanks for the help you gave us in this matter. A bitter grief has been taken from us by this news. We consider your aid in this really arduous task of getting our parents out of Poland a new proof of your sympathy also for our private needs."

Rocket Team II

In late June 1947, after the Germans had been in the United States for two years, the matter of their remaining in America and becoming citizens was raised. Representative John Dingell (D.-Mich.) said, "I have never thought we were so poor mentally in this country that we have to import those Nazi killers to help us prepare for the defense of our country."*

Keeping the Germans on American soil was not only a means of building U.S. rocket power; it was also seen as an element of national defense, as once-imagined fears of Soviet power became real. The last thing America wanted was to see the Rocket Team disappear behind the Iron Curtain.

The V-2s that had been shipped from Germany were sent to the Ordnance Proving Ground at White Sands, New Mexico. The facility was near Fort Bliss and close to where Robert Goddard had conducted his work. At Fort Bliss, the team assembled, tested, and launched V-2s. In the process, they taught Americans the intricacies of handling large rockets and missiles. On April 16, 1946, the Army launched the first reconstructed V-2, and during the next six years, sixty-four more of them were launched from White Sands. Instead of carrying explosives, the nose cones held instruments, which were usually recovered by parachute.

Using the salvaged rockets and new off-the-shelf hardware, American scientists were able to test the frontiers of space, probing the upper reaches of the atmosphere in 1946. Many of these early rockets carried Geiger counters and other radiation detectors to high altitudes. James A. Van Allen, S. Fred Singer, and other pioneers of space science wanted to see how the intensity and nature of cosmic rays varied with altitude and latitude. Rocket number 13 was launched at White Sands in October 1946 and took motion pictures of the Earth from an altitude of sixty-five miles, suggesting that space was an ideal platform for military reconnaissance. A mosaic was created from the film, covering 800,000 square miles.

The value of the rockets was immense, and reporters flocked to New

*This is the father of Representative John Dingell Jr., elected in 1955.

Mexico to report on the utility of a former enemy weapon. "Thanks to the V-2s . . . more had been learned about the upper atmosphere in the last two years than in the previous twenty," Daniel Lang wrote in the *New Yorker* in 1948 after a trip to White Sands. Sensors in a V-2 revealed, among other things, that the average distance between molecules in the upper atmosphere is 370 inches, while the distance on the surface of the Earth is only about one millionth of an inch. Late in 1946, V-2 sensors discovered the Earth's ozone layer.

Although wireless telemetry had already been demonstrated in 1925 by a Russian scientist named Pyotr A. Molchanov transmitting from a radio in a balloon, the art had become advanced with the V-2 experiments. Scientists on the ground knew exactly what was happening with their instruments at all heights. The ability to link Earth and space electronically was being established and redefined: Data were transmitted back on solar spectroscopy, radiation, meteorites, temperatures, winds, and ambient pressures.

American scientists became so intrigued with the possibilities of high-altitude research that the White Sands group was flooded with requests that instrumentation packages be taken aloft. This clamor finally had to be coordinated through James A. Van Allen and his Upper Atmospheric Rocket Research Panel.

There were also experiments with animals, which through trial and error began to pave the dangerous path to manned spaceflight. A V-2 designated Albert, in honor of its passenger, a monkey, was launched in June 1948. The monkey, the first American primate in space, died of suffocation. In June 1949, Albert II was launched into space but died on impact.* Despite the outcome of these flights, they proved that primates successfully survived the rigors of flight, including the stress of withstanding more than five times the force of gravity (5.5 Gs), which made their weight increase from nine pounds to forty-nine and a half pounds during liftoff acceleration.

There were military experiments too. In 1947, for example, the Rocket Team proved to the Navy that a large missile could be launched from an aircraft carrier by staging a demonstration from the deck of the USS *Midway*.

The Rocket Team proved invaluable as a hands-on graduate school in

*During 1949 two more V-2 flights of this type were conducted, and in each case, the monkey survived the flight but succumbed before his capsule could be located. The first successful recovery of animals from a rocket flight in the West did not occur until 1951, when a monkey and eleven mice survived the launch of a Navy Aerobee rocket.

the basics of rocketry for the Army, academia, and American industry. It was later estimated that the team enabled the U.S. space program to gain ten years and save billions of dollars in research-and-development time. Although there was never any question that the Germans would be allowed to remain indefinitely, by 1949 the "temporary" status of the Germans on American soil was reversed when they were offered permanent employment, albeit surreptitiously. The Germans were all herded into El Paso city buses, taken across the border to Mexico, issued visas, and returned legally to Texas. Later, when they applied for U.S. citizenship, they would write down under point of entry, "El Paso City Line."

Above all, America did not want to lose von Braun, who in 1947 returned to Germany to marry his eighteen-year-old cousin, Maria Louise von Quistorp. Soon after the wedding, he and his wife returned to the States. American intelligence had been present at the wedding and short honeymoon, fearing he would be grabbed and taken behind the Iron Curtain. But the Russians already were discovering their own counterpart to Wernher von Braun.

Soviet Advances in Rocket Technology

The Soviet Union wanted the V-2 technology as much as, or more than, the United States and decided that it would bring home the means of production. Sergei P. Korolev, a Red Army colonel who saw the German rocket as the key to his childhood dreams of space travel, had been sent to Germany on September 8, 1945, to study and develop the V-2. He was ultimately part of the group responsible for transferring V-2 production and testing facilities to the USSR.

Korolev was officially known as the "chief designer" during his days of glory; only a handful of people behind the Iron Curtain knew his true identity. He was kept anonymous by an ideology that wanted his work to be seen as the product of a system of committees and collectives, rather than of individual genius; the Soviets also feared that if his identity were known, he would be assassinated by the U.S. Central Intelligence Agency. His name was not known to American intelligence until a year before his death and was not revealed publicly until after his death.*

*The world would soon become obsessed with trying to determine who this man was. The first rumor that he was the "chief designer" came after Nikita Khrushchev mentioned him by name in a toast proposed at the marriage of cosmonaut Valentina Tereshkova, the first woman in space, to cosmonaut Andrian Nikolayev.

Model of Russian GIRD rocket, which was test-launched on August 17, 1933, under the leadership of Sergei Korolev. (Tass/Sovfoto)

Korolev was the individual responsible for the Sputniks, Luniks, and cosmonaut Yuri Gagarin's pioneering 1961 venture into space. He built the Vostok, Voskhod, and Soyuz spacecraft, which carried all of the Soviet cosmonauts into orbit. Trained in aeronautical engineering at the Kiev Polytechnic Institute, he cofounded a Moscow rocketry organization by the name of Group for Investigation of Reactive Motion (in English translation). Like the VfR in Germany, and Robert H. Goddard in the United States, the Russian organizations in the early 1930s were testing liquid-fueled rockets of increasing size. Korolev's book, *Rocket Flight in the Stratosphere*, was published in 1934 by the USSR Ministry of Defense.

A curtain of deep secrecy fell over Russian rocket activities, not to rise again until the launching of Sputnik revealed just how much the Soviet Union had accomplished in the intervening quarter century. Soviet research in the field of rocketry was concentrated in Scientific Research Institute 3, known as NII-3 and led by Ivan Kleimenov.

NII-3 was wiped out during one of Joseph Stalin's purges. Kleimenov and his deputy, Georgiy Langemak, were arrested and executed in 1937. Stalin believed that the institute was at the heart of a cabal preparing to use the new rocket technology to overthrow the government. In addition, because Kleimenov had once worked in Berlin for Aeroflot, the Soviet airline, Stalin saw this as de facto proof that he was a German spy.

Cosmonaut Yuri Gagarin (left), the first person to fly in space, with Sergei P. Korolev, head of the Russian space program. (Sovfoto)

In the early morning of June 27, 1938, four men—two members of the secret police and two "witnesses"—entered Korolev's apartment and dragged him away from his wife and three-year-old daughter without so much as a change of underwear. He was beaten and forced to confess to trumped-up charges of sabotage and, without benefit of trial, sentenced to ten years in prison.

Shuttled at first from one grim prison to another, at some point in 1939 he landed in what his biographer, James Harford, called "one of the most dreaded of all prisons, a camp in the Kolyma area of far eastern Siberia, part of the notorious gulag system (gulag is the acronym for Main Directorate for Corrective Labor Camps), made famous in the West by the publication of Aleksandr Solzhenitsyn's *The Gulag Archipelago*." Conditions in this prison—where 10 percent of the population died annually from tuberculosis, malnutrition, or execution—were so bad that Korolev lost all of his teeth and acquired a heart condition, which contributed to his early death.

At the beginning of World War II, some of the most brilliant imprisoned engineers were allowed to form a *sharashka*, a design bureau in prison to contribute to military technology. Korolev's transfer from the gulag to such a group in 1942 quite possibly saved his life. By 1944, this behind-bars collective successfully completed testing of the liquid-fuel rocket engine RD-1, which was accepted for mass production and installation on a number of Soviet fighters and bombers. This work was personally approved by Stalin, who, on July 27, 1944, acquitted and removed convictions from Sergei Korolev and thirty-four other engineers. Korolev was now released. According to his official file, he had undergone "full rehabilitation," a phrase that Cold War historian Jeffrey Robinson says was "considered a tacit recognition of a wrongful sentence."

Korolev had been in jail for six years, four of which were lost forever to the Soviet effort in space. In August 1945, he was finally able to make two short visits to Moscow to visit with his family and friends. (His wife had divorced him during his absence.) An old associate and space visionary, Mikhail Tikhonravov, who had seen a captured V-2, raved about it to Korolev. "You can't imagine how big this pot is," said an awed Tikhonravov, referring to the V-2's combustion chamber.

A month later Korolev went to Germany as part of a team collecting spoils of the V-2 program. With the help of German engineers, during late 1945 and all of 1946, the Soviets reestablished a V-2 production line in East Germany, turning out about a dozen missiles. Enough spare parts to create another were sent directly to the Soviet Union.

Korolev and his fellow engineers started to consider their own and German ideas on an improved V-2, recast as a Soviet rocket. Two special trains fully equipped with test labs, living compartments, and missile carriers were secretly outfitted to support V-2 test launches in the USSR. In October 1946, the best German engineers recruited by the Soviets were ordered on the trains and sent to the various locations in the USSR to assist in the organization of missile production and design. By the beginning of 1947, Soviets had completed the transfer of all work on rocket technology from Germany to remote locations in the USSR.

In the autumn of 1947, the Soviets began to explore the upper atmosphere with rockets of the V-2 class; from 1949 on, they used Pobeda rockets, which were greatly improved versions of the V-2. (At this time, the United States was still using its stock of original V-2s.)

The Soviets also started to explore the possibility of an Earth-circling satellite. After all, in 1903 Konstantin Tsiolkovsky had shown mathematically that a device launched at a certain velocity would achieve Earth orbit. Forty-five years later, Tikhonravov made the case to Korolev for developing just such a device and reasoned that if the major obstacle to launching a satellite past the atmosphere of the Earth and into orbit was nothing more than a matter of thrust, all that was needed was raw rocket power. To obtain sufficient power, Tikhonravov proposed strapping several rockets together in a cluster: a bouquet of five rockets in the first stage, three in the second, and one in the third.

When Tikhonravov submitted a report on this topic to the Academy of Artillery Sciences, which was planning its annual meeting for 1948, it was treated with official skepticism. Academy president Anatoli Blagonravov, concerned about Stalin's reaction, said, according to Korolev's biographer, James Harford: "The topic is interesting. But we cannot include your report. Nobody would understand why. . . . They would accuse us of getting involved in things we do not need to get involved in." However, Harford noted that "what Blagonravov said officially was not what he thought instinctively. This courteous, mild-mannered, chain-smoking, white-haired former general, who would become in later years one of the chief spokesmen for the Soviet Union in the United Nations Committee on the Peaceful Uses of Outer Space, was bothered by the wary reception of Tikhonravov's ideas." Risking the derision of his colleagues, Blagonravov put the report on the agenda, thereby giving Tikhonravov—and Korolev—license to study possible satellite designs.

Dead Ends and Detours on the U.S. Path to Space

Although the U.S. Army had absolutely no interest in such things, in the late 1940s von Braun became increasingly interested in Earth-circling satellites, space stations, and manned probes. He embarked on an ad hoc career to which he would devote himself for the rest of his life: the selling of outer space.

He was a born salesman in a culture that was highly receptive to sales pitches, even when the product was a bit odd. He would saunter into Rotary and Kiwanis clubs to talk about space to audiences who, though startled by his message, saw him as one of their own: a booster and a backslapper. He loved reporters, and cameras were kind to him. He always had time for anyone who bought ink by the barrel. Journalists passing through V-2 country were given the full story of humans' destiny in space. He told them that within a decade it would be possible to have a space station in orbit, which could eventually be used as a refueling station for a voyage to the Moon.

During the quiet southwestern nights at Fort Bliss, von Braun began putting his dreams on paper. In 1947 he finished writing a book about a manned trip to Mars. *The Mars Project* was turned down by no fewer than eighteen American publishers before it was finally published five years later in German.

Even though von Braun had trouble finding a publisher, space was becoming a hot topic. The public of the postwar period was being treated to an increasingly steady diet of articles on the subject. One of the most spectacular appeared in the July 23, 1945, issue of *Life* and contained detailed drawings of a large, manned space station, including a large space mirror, envisioned by the Germans at Peenemünde. The American public now realized how close the Nazis had come to entering space.

Although small groups of devotees in the United States as far back as the mid-1940s had made serious proposals to launch a spacecraft into Earth orbit, they were outside of government and lacked power. The U.S. government did not begin thinking seriously about space until 1950.

The U.S. path to space was a tortured one, filled with dead ends and detours. For the U.S. military, the immediate postwar period was one of both opportunity and constraints. New technology had come out of the war—atomic weapons, jet aircraft, missiles, and more. But with the end of the war, the nation demobilized, and defense spending plummeted—from $81.5 billion in 1945 to $13.1 billion in 1947, leading the way to unrelenting interservice rivalry and inertia. With the budget cuts, the branches of the military had a hard time redefining themselves, their purpose and territory.

In October of 1945—the same year the German Rocket Team was allowed into the United States—the Navy commissioned a special committee report on the feasibility of rocketry, which recommended, among other things, the development of an instrumented Earth satellite. Fascinating options were presented, but nothing ever happened due to lack of interest and money. Robert P. Haviland of the Navy's Bureau of Aeronautics, inspired by a report on space rockets captured with the Germans, then wrote a memo to the committee proposing satellite development by the Navy, but it was, to use his word, "scorned."

At the end of the war, the U.S. Army Air Corps did ask the major aircraft companies to submit secret proposals for the design of an Earth-orbiting satellite. Douglas Aircraft won the contest and assigned the work to its newly formed Project RAND (for Research and Development), an arm of Douglas located in Santa Monica, California. In its May 1946 report, prepared by a team of fifty researchers, Project RAND envisioned a 1951 launch date to be feasible, with communications, weather, and spy satellites following in the wake of the first demonstration satellite.

The RAND analysts estimated that the United States could design, construct, and launch a satellite for about $150 million. The report laying out the course of an orderly American progression into manned spaceflight concluded that "since mastery of the elements is a reliable index of material progress, the nation which first makes significant achievements in space travel will be acknowledged as the world leader in both military and scientific techniques. To visualize the impact on the world, one can imagine the consternation and admiration that would be felt here if the U.S. were to discover suddenly that some other nation had already put up a successful satellite."

The authors of the RAND report admitted that although their crystal ball was a bit cloudy, two things seemed clear. First, "a satellite vehicle with appropriate instrumentation can be expected to be one of the most potent scientific tools of the Twentieth Century." Second, the "achievement of a satellite craft by the United States would inflame the imagination of mankind, and would probably produce repercussions in the world comparable to the explosion of the atomic bomb."*

The top brass was unimpressed, and, for all intents and purposes, the study was shelved. It described the state of the art rather than reflecting the Army Air Corps' interest in space. The recommendations of the RAND study were read and circulated in military and scientific circles, of course, but not implemented even though more than theory was at work in at least one area: space communications. A few months before, the Army Signal Corps announced that on January 25, 1946, it had bounced radio signals off of the Moon and received the reflected signals back on Earth. This did not provide an effective communications link, but it proved that radio transmissions through space and back to Earth were possible with a moderate amount of power. These signals could be used to control manned or unmanned spacecraft.

During the same month that the Project RAND report appeared, another issue was brought into focus. The War Department Equipment Board—popularly referred to as the Stillwell Board because it was chaired by Army general Joseph W. "Vinegar Joe" Stillwell—warned that the United States must continue to fund research and development to ensure that the nation would not be technologically surprised in the future. But this suggestion was also rejected.

In August 1947, the National Security Act established the Department of Defense, uniting the military under the leadership of a new cabinet-level

*This report made headlines in the weeks after Sputnik's launch because it was one of the classified documents that were turned over to the Russians by atomic spies Ethel and Julius Rosenberg. The incident, which will be discussed in chapter 7, is noted here because of the irony that this call to space may have been heard by the Russians.

post—the civilian secretary of defense. Congress saw the new department as a coordinating force for the armed forces. But a month later, the equation changed as the Air Force was created as a separate service, with resources coming primarily from the Army Air Corps. The Army and Navy now had a brash new competitor for weapons and funds. Even before the creation of the U.S. Air Force, the nascent missile efforts of the Army Air Corps were already hard hit by the budget cutbacks—in December 1946, the AACF's budget for missile research and development was cut by more than 50 percent, from $29 million to $13 million.

In the wake of the RAND report, satellites were suddenly important and worth fighting over. In October 1947, the Committee on Guided Missiles of the Joint Research and Development Board was given authorization to evaluate satellite studies produced within the various branches of the armed services. The committee recommended, however, that satellite development programs *not* be endorsed until a definite and compelling military use for them was determined.

Despite this, the Navy began to explore the idea of a vehicle capable of putting a satellite in orbit and decided to make a bold move by claiming jurisdiction over the development of defense-related satellites. This set off some jockeying for position. A few weeks later, the Air Force's vice chief of staff, Hoyt S. Vandenberg, announced that the Air Force would only attempt to develop satellites "at the proper time." Vandenberg, however, stated strongly that any type of satellite was a logical extension of strategic airpower and, therefore, should be the responsibility of the Air Force. The next day the Navy withdrew its claim for control of satellite development just as suddenly as it had made it. In September 1948, the Committee on Guided Missiles of the Joint Research and Development Board gave official authorization for the Army to begin feasibility studies on the launching of satellites.

At the end of 1948, Secretary of Defense James V. Forrestal in his *First Report of the Secretary of Defense* publicly announced that all three branches of the service were now studying the issue. The announcement created a public outcry, for Americans equated such activity with the squandering of taxpayers' money. As a result, a fledgling Navy plan to create a high-altitude test vehicle (HATV), already being designed by the Glenn L. Martin Company, was brought to a halt. It was later claimed that this effort was, at the time of its termination, ahead of the Soviet Union in the areas of propulsion and structural engineering. The Navy would have launched its satellite in 1951.

The military services, especially the U.S. Air Force, were struggling with conflicting notions of how the country should move into the age of jet aircraft, missiles, and rockets. In the context of this struggle, the Air Force saw a need for a stable, highly skilled crew of analysts to help with the evaluation of the various alternatives. Friction had been developing between the structured order at Douglas Aircraft (where there were policies regarding where and when coffee could be consumed) and the freewheeling, academic thinkers of Project RAND. A staff member wrote about one of the many problems: "The matter of hours of work were—and are yet—a substantial trial. Academic people have irregular habits and have never taken kindly to the eight-to-five routine. We had one man who rarely showed up before two o'clock, and another who almost never went home—this, mind you, in an organization where they physically locked the doors at about five, and kept them locked until about eight in the morning." This policy was changed so that RAND workers could work nights and weekends, but the problem underscored the conflict between an aircraft company and a group that had its own philosopher (who had to be described as Design Specialist A to fit the Douglas job description).

To effect what amounted to an amicable divorce, money was raised—including a crucial $100,000 donation from the then-nascent Ford Foundation—and in 1948 Project RAND was taken from Douglas and established as the RAND Corporation, a civilian "think tank" to which the Air Force could turn for independent analysis. According to its Articles of Incorporation, it was formed "to further and promote scientific, educational and charitable purposes all for the public welfare and Security of the United States of America." Although the 1946 RAND satellite proposal had been turned down, it represented the kind of clear, forward-looking thinking the Air Force was looking for.

Early in 1949, the Air Force asked RAND to prepare more studies on the potential usefulness of space. The next space report, delivered two years later, would have a major impact on America's plans and policies and ultimately affect the outcome of the Cold War.

The U.S. Space Program Moves East

For Wernher von Braun and the Rocket Team, the culmination of their Fort Bliss–White Sands years was the creation of the key ingredient for entering space: the development of multistage rocketry. In 1948 one of the WAC-Corporal missiles from the Jet Propulsion Laboratory was stacked on top of a V-2 to create a two-stage rocket named Bumper. Firing first to provide the initial liftoff, the V-2 boosted both into the sky and then dropped away as the WAC-Corporal ignited to carry its load to greater heights than ever before.

The following year, a V-2 modified to carry a WAC-Corporal second-stage rocket (known as Bumper 5) was launched from White Sands. The second stage reached a record altitude of 244 miles, becoming the first man-made object to reach that far into space. The next day in Washington, the Army's ordnance research chief, Major General Henry B. Sayler, said that the performance at White Sands "brings nearer the day when it may be possible for man to launch a satellite 'rocket' that can circle the Earth continuously."

But dreams of orbiting satellites were quickly overshadowed in September of 1949, when an American weather plane in the Pacific detected radioactive particles in the atmosphere, indicating that the Soviets had tested their own atomic bomb in late August. This discovery had a shocking effect on the American political mood, and a few weeks later President Truman ordered the development of a much more powerful thermonuclear bomb—the hydrogen bomb. The arms race that would last for decades and absorb trillions of dollars and rubles had begun in earnest.

The Soviet Union was quick to grasp the potential of an atomic bomb attached to a long-range missile; well before they had the bomb, the Soviets had begun working on gargantuan rocket clusters able to cross oceans. (In the United States this approach was initially considered impractical, and the American military chose instead to rely on the long-range bomber as its prime offensive weapon.)

In March 1947 a state commission was created to report on the feasibility of producing long-range ballistic missiles. Its first recommendation was that one of its Russian-made V-2s be refined and improved into a more effective weapons system. At the end of October, the first improved Soviet version of a German V-2 was launched from Kazakhstan. The vehicle continued to be refined and was eventually deployed as the Pobeda mobile missile, which had a range of 500 miles. Following development of the Pobeda, all of the German production engineers, technicians, and workers remaining in the Soviet Union were repatriated to Germany.

As a result of the Soviet nuclear detonation, missiles took on new importance for the United States. In October 1949, the Army decided to move von Braun and his team to the Redstone Arsenal in Huntsville, Alabama, where they would develop ground-to-ground ballistic missiles for nuclear war, uniting rockets based on proven V-2 technology with warheads based on the tech-

The first Cape Canaveral firing of a Bumper Project rocket, on July 24, 1950. (Patrick Air Force Base, Florida)

nology America had used against Japan. Although the actual move would not take place until April 1950, it was clear to the Germans and the Americans who were now part of the team that they were full-fledged employees of the U.S. Army, and, as had been the case when the Germans had been moved to Peenemünde, their job was now to create weapons, not spaceships. Their first major assignment was to develop the new Redstone rocket as a tactical ballistic missile. The purpose of the missile as it developed would be to direct four-megaton missiles into the Soviet Union from bases in West Germany. The V-2 was returning to Germany painted in American colors.

The Germans, sick of the treeless desert and the confinement and restriction of a place cruelly named Bliss, immediately fell in love with Huntsville. This green, leafy place—"The Watercress Capital of the World," as it called itself—reminded most of them of Germany in the warmer months. Huntsville also represented personal freedom, a chance to build homes and put Hitler, the SS, and the war behind them. They could take out library cards, buy land, build houses, raise families, and become citizens. Few of them ever left. Huntsville was to remain von Braun's home for the next twenty years, a period in which the city's population increased tenfold.

Created in 1941 for the production of chemical compounds and small solid-fuel rockets, the arsenal had everything the team needed: laboratories, assembly areas, and access to a firing range. Huntsville was not a place for firing rockets, but booster rockets could be moved on large barges on the Tennessee River which ran to the Atlantic Ocean, then along the coast to the new rocket-launching area at Cape Canaveral on the Banana River in Florida. As the need for more room to fire the rockets became evident, the

Joint Long Range Proving Ground had been established at remote, sparsely inhabited Cape Canaveral, Florida, in 1949. Even the biggest rockets imaginable could be sent to the cape in pieces on oceangoing barges.

The transfer of launch operations to Florida coincided with the transfer of the Army's missile program from New Mexico to Alabama. On July 24, 1950, a two-stage Bumper 8 was launched from the cape's new launchpad 3 at 9:28 A.M., eastern time. It was the first rocket launched from the cape. The purpose of the mission was to test methods of stage separation while a rocket was performing a near-horizontal flight. At first-stage separation, it suffered a structural failure at the point where the first stage is mated to the second, throwing the rocket radically off course. The rocket was destroyed by the range safety officer, who punched a button and sent it crashing into the Atlantic Ocean about forty-eight miles downrange. It was a poignant moment, symbolizing the end of an old era—the first stage of the rocket was one of the last V-2s made in Germany.

The America of 1950 was a fearful place in the grips of a Cold War with the Soviet Union, which had quickly spread to other parts of the world. The situation had become hotter on June 25, 1950, when forces of the Democratic People's Republic of Korea poured across the thirty-eighth parallel into South Korea. Communist China soon joined forces against the United States and South Korea. As a result of this and other Cold War tensions, the policy makers in Washington were beginning to fully realize the tremendous potential of missiles as military weapons and were funding more experimental programs. The workload in Huntsville grew, and by June 1951, more than 5,000 people were on the payroll at the arsenal. Huntsville's population doubled.

The von Braun team worked to develop the Redstone, essentially a super-V-2 rocket named for the U.S. Army arsenal where it was being designed, and later, a variety of medium- and long-range intercontinental ballistic missiles. These became the starting point of the U.S. space program. Ironically, missiles such as the Redstone, Jupiter, Atlas, and Titan, which would eventually launch satellites and astronauts into space, were born as weapons of Armageddon, conceived to destroy cities and kill millions.

CHAPTER FOUR

An Open Sky

> Mankind's first great step forward into outer space consists in flying beyond the atmosphere and creating a satellite of the earth. The rest is comparatively easy, even escape from our solar system.
>
> —Konstantin Eduardovich Tsiolkovsky, 1926

America's journey into space can arguably be traced to a gathering at James Van Allen's house in Silver Spring, Maryland, on April 5, 1950. The guest of honor was the eminent British geophysicist Sydney Chapman, who was in the Washington area for a brief visit. The other guests were S. Fred Singer of the University of Maryland; J. Wallace Joyce, a Navy geophysicist and State Department adviser; Lloyd V. Berkner, head of the new Brookhaven National Laboratory on Long Island; and Ernest H. Vestine of the Department of Terrestrial Magnetism of the Carnegie Institution.

After dinner, talk turned to the question of how to obtain simultaneous measurements and observations of the Earth and the upper atmosphere from a distance high above the Earth. Berkner suggested that perhaps it was time to stage another International Polar Year, a global extravaganza of science.

The first International Polar Year had established the precedent of global scientific cooperation in 1882, when scientists from a score of nations pooled their efforts for a year in studying polar conditions. A second International Polar Year took place in 1932, and it was always assumed the third would be scheduled fifty years later, in 1982.

Berkner's proposal to shorten the interval from fifty to twenty-five years made sense because astronomers already knew that 1957–58 would be a period of maximum solar activity. The plan was to unite scientists from the United States, the Eastern Bloc, and other nations to conduct atmospheric studies on an international scale. Valuable data would be collected, and at the same time the event would demonstrate the world's common scientific values at a time of discord and Cold War.

The Silver Spring group also hoped the International Polar Year would be anchored by one great, headline-making event: the launching of the world's first artificial satellite. The idea emerged full-blown from the gathering that nations could be encouraged to convert ballistic-missile technology to peaceful uses by launching an orbiting science lab. Two months after the Silver Spring meeting, the National Science Foundation, a government agency whose main function is to support basic research by granting money to American universities, began to encourage satellite proposals. Whereas the mission of the Army, Navy, and Air Force was national defense, that of the foundation was the fostering of scientific discovery—this was the difference between applied and basic research.

Suddenly, proposals started making the rounds as Berkner and Chapman, along with thirty-seven European scientists, developed the International Polar Year idea further, getting it in shape to present to the International Council of Scientific Unions. In 1952 the council appointed a committee to make arrangements; extended the scope of the study to the whole Earth, not just the polar regions; and fixed the duration of the "year" at eighteen months—July 1, 1957, to December 31, 1958—to embrace the major periods of solar activity. The council renamed the undertaking the International Geophysical Year (IGY). It eventually embraced sixty-seven nations, and in October 1954, the council, including delegates from both the United States and the USSR, officially adopted a resolution calling for artificial satellites to be launched during the period to map the Earth's surface.

As the IGY idea developed and the word *satellite* began to take on a new meaning, popular interest in space grew. Von Braun had by now become a vocal and public advocate for space, joining a small but growing movement that included other strong voices. In 1951 Sir Arthur C. Clarke's *Exploration of Space* was a Book-of-the-Month Club selection, and Willy Ley's *Rockets, Missiles, and Space Travel* was already a perpetual best-seller. New York's Hayden Planetarium and the Air Force both staged well-publicized symposia on space, inspiring stunningly illustrated articles on outer space in the popular American weekly magazines.

On March 22, 1952, *Collier's* magazine published "Man Will Conquer Space Soon." The main contributor was Wernher von Braun. He was assisted by Joseph Kaplan, professor of physics at UCLA; Fred L. Whipple, chairman of the astronomy department at Harvard; Heinz Haber of the U.S. Air Force Department of Space Medicine; Willy Ley, writer and space authority; and Cornelius Ryan, senior writer at *Collier's*. A team of illustrators, including Chesley Bonestell and Fred Freeman, was assembled to bring the ideas to life. Over the next couple of years eight articles by this team were published in *Collier's*—one of the nation's largest magazines, with an estimated pass-around circulation of 12–15 million readers.

Von Braun and company described his concept of a space station for *Collier's* readers. This station would have a diameter of 250 feet, travel in a 1,075-mile-high orbit, and spin to provide artificial gravity. In his vision, it would be the perfect jumping-off point for lunar expeditions. The space station and the Moon voyages would be serviced by a fleet of reusable three-stage space vehicles—an idea strikingly similar to the space shuttle. The last installment of the series called for a flotilla of ten spacecraft to be built in space, then push toward Mars on an eight-month journey. Once in Mars's orbit, a manned "landing boat" equipped with skis would glide down to the planet's polar icecap to establish a landing strip for two-wheeled boats.

Von Braun also worked with the Walt Disney studios as a technical director for three popular television films about space exploration based on the *Collier's* series.*

The clear intent of von Braun and his fellow space enthusiasts was to build a constituency for outer space through greater public interest in the future of the space program, a program which, in fact, did not come into existence until 1958. *Time* magazine's editors disapproved of these efforts, noting that an "oversold public . . . happily mixing fact and fiction, apparently believes that spaceflight is just around the corner."†

Von Braun was also trying to sell the idea of putting a satellite into orbit as soon as possible. One public proposal, which was aggressively marketed,

*The Disney films resurfaced on the cable Disney Channel and can still be seen today.

†In 1993 "Blueprint for Space," an exhibit on the *Collier's* articles, was staged at the Smithsonian Institution's National Air and Space Museum in Washington. Mary Henderson, curator of that exhibit, remarked at the time of its opening: "People saw in the stories that space travel could be real. They saw it could happen and that it was worth supporting with tax dollars." Speaking of the collective impact of the magazine and the Disney films, Henderson said, "A generation grew up with this stuff. They read books, saw movies, played with space toys. It got them ready for what was to come."

had a number of names, including "American star." By 1953, when Jonathan Norton Leonard, science editor of *Time* magazine, published the book *Flight into Space*, von Braun was in full swing. Leonard's book ended with a plea from von Braun for a display. "Even a small rocket," he told Leonard, "can be made to shine at dusk against a dark sky. Once established in its orbit, it will inflate a white plastic balloon 100 feet in diameter. Swinging swiftly around the earth 200 miles above the surface, the balloon will gleam as brightly in the sunlight as a first magnitude star." Von Braun claimed it would cost little and "make an enormous impression on the people of Asia."

Eye in the Sky

While the IGY satellites were to be international and public, top strategists in the U.S. government began to consider the clandestine use of satellites as "eyes in the sky." On October 4, 1950, a second secret report was issued by the now-independent RAND Corporation. It went beyond the question of whether a satellite could be launched, and explored the conditions under which it should be launched. The RAND report stressed the importance of the satellite as an instrument of reconnaissance—spying. The Soviet Union was cloaked in a veil of extreme secrecy, and America's leaders desperately needed to know what was going on in terms of missile deployment, troop movement, and naval activity.

According to historian Walter A. McDougall in his 1985 book *The Heavens and the Earth*, the second RAND report emphasized that the United States should not provoke Russia with U.S. military satellites overhead; therefore RAND recommended that the first U.S. satellite be billed as "experimental" and be launched on an equatorial orbit, avoiding the USSR altogether. RAND held that this would help establish the concept of "the freedom of space."

A follow-up RAND study, which appeared six months later, went into much greater detail on how all of this would work by using television technology in satellites for intelligence and weather reconnaissance. The study also suggested that detailed high-altitude observation of the Soviet Union might be feasible. To confirm and further the RAND findings, the Air Force created and convened the "Beacon Hill" study group in January 1952 to assay strategic aerial reconnaissance under the auspices of Project Lincoln at the Massachusetts Institute of Technology. The potential value of reconnaissance satellites was soon confirmed by some of the leading scientific and technical thinkers in the country, including Edwin Land, inventor of the Polaroid camera.

As weapons of pure destructive power became more dangerous, the need for international intelligence grew more crucial. In November 1952 the United States detonated the first fusion, or hydrogen, bomb. It was exponentially more powerful than the atomic bomb, with the equivalent force of ten megatons—or 10 million tons of TNT. The Hiroshima atomic bomb had an equivalent force of 12,500 tons of TNT.

President Dwight Eisenhower was interested in controlling the use of nuclear weapons and bringing order to American defense. Upon taking office in 1953, he set his staff to work on a "new look" for defense. Space historian Eugene M. Emme distilled the assumptions at work in the "new look" to these four points: (1) The United States would not start a war; (2) nuclear war was a "catastrophe beyond belief"; (3) American armed forces should be developed and designed to deter nuclear war; and (4) military strength was only one factor in sustaining genuine national security for the "long haul," along with national will, intellectual genius and vigor, and fiscal solvency.

In August 1953 the Soviets tested their first prototype hydrogen bomb, and later in the year U.S. intelligence sources revealed that the Soviet intercontinental ballistic missile (ICBM) program was well on its way to becoming reality. The assessment was that the Soviets were significantly ahead of the United States. They not only had nuclear weapons but also had the means to deliver them against the continental United States. This was powerful validation of RAND's call for reconnaissance satellites to keep Americans aware of what was going on in the Soviet Union.

Also in 1953, the United States successfully developed a high-yield, lightweight thermonuclear weapon. This breakthrough, combined with the intelligence report that the Soviets were developing long-range ballistic missiles, led President Eisenhower to move ahead with the development of an intercontinental ballistic missile that could carry U.S. lightweight warheads across oceans. The three branches of the military services each wanted to be given the task of running the missile program.

In February 1954 the stakes were raised as Eisenhower quietly ordered that the highest possible priority in the use of talent, money, and material be assigned to the intercontinental ballistic missile. As this went ahead and nuclear testing continued, he began looking for a way to thaw the Cold War and approach the issue of nuclear disarmament. After the death of the intractable Joseph Stalin in March 1953, Eisenhower began hunting for an opening to talk directly with Nikita Khrushchev—he even offered to go to Moscow. There were reasons to hope for a break in hostilities, especially as the IGY grew nearer and as four nations—the United States, the USSR, France, and England—agreed to meet in July 1955 at a summit in Geneva.

At an informal dinner with Khrushchev and other Russian leaders on the eve of the summit, President Eisenhower said in a voice loud enough so that the whole table could hear, "It is essential that we find some way of controlling the threat of the thermonuclear bomb. You know we both have enough weapons to wipe out the entire Northern Hemisphere from fall-out alone." The remark was recorded by John Eisenhower, the president's son, who was told after the dinner by Nelson Rockefeller, special White House assistant for psychological warfare, that his father had "a bombshell" that he would drop during the formal session.

On July 21, 1955, Eisenhower presented his "Open Skies" proposal. It was an audacious and unexpected attempt to exert control over nuclear mobilization and surprise attack. Under the plan, each nation would not only offer the other a blueprint of its military sites but provide "ample facilities for aerial reconnaissance, where you can make all the pictures you choose, and take them back to your own country for study." Open Skies, Eisenhower reasoned, would not just prevent surprise attacks but would be the beginning of "effective inspection and disarmament."

The Russians disliked the idea. Soviet prime minister Nicolai Bulganin objected to Open Skies for several reasons. According to Eisenhower, Bulganin reasoned that "if the military in both the United States and the Soviet Union were to learn all about each other's armed forces, they would automatically begin agitation for increases in their own; therefore the Open Skies would increase rather than decrease armaments."*

Khrushchev later told Eisenhower: "That is a very bad idea; it is nothing more than a spy system." He was right in the sense that it was an open, mutually beneficial spy system.

Nevertheless, the time was optimal for the United States to make public an important decision that had been kept secret: On July 29, eight days after the Open Skies proposal, White House press secretary James Hagerty announced that the United States would launch "small unmanned Earth-

*In 1989 President George Bush incorporated the Open Skies concept in a treaty proposal. It was signed by twenty-five nations on March 24, 1992.

circling satellites" as part of IGY. The White House announcement was a source of elation to American scientists, who after many years of waiting had won support for the project that would give them a revolutionary new tool for basic research beyond the Earth's atmosphere.

This first public announcement of America's decision to go into space was big news in the world press. The following day, the Soviet government made an announcement of its own, affirming its commitment to launch Soviet satellites during the IGY. Three days later, at the sixth International Astronautical Congress in Copenhagen, Leonid I. Sedov, head of the Soviet delegation and the newly appointed chairman of an Academy of Sciences Commission on Interplanetary Communications, first alluded to the American satellite. Choosing his words carefully, he said: "From a technical point of view, it is possible to create a satellite of larger dimensions than that reported in the newspapers which we had the opportunity of scanning today." He then added, "The realization of the Soviet project can be expected in the comparatively near future. I won't take it upon myself to name the date more precisely."*

Project Slug

When the White House made the decision to announce a series of IGY satellites, nothing was said of who would build them—only that proposals would be sought from various agencies of government. There had already been a great deal of jockeying for position, with Wernher von Braun at the front of the line.

A year earlier, Frederick C. Durant III, president of the International Astronautical Federation, had requested that the Office of Naval Research (ONR), a group heavily staffed with civilian scientists, set up a meeting to investigate the possibility of having the Army and Navy cooperate in the launching of a satellite. The meeting was set for June 25, 1954, when Wernher von Braun was expected to be in Washington. A joint Army-Navy team secretly met at the Washington office of the ONR, where von Braun presented a detailed Army proposal for the launching of satellites using existing Redstone missiles mated to upper-stage clusters of slender antiaircraft Loki

*But Sedov, it seems, was to a certain degree going out on a limb, since no official decision had yet been made that there would be a Soviet satellite in IGY. In fact, not until January 30, 1956, would the Council of Ministers issue a decree authorizing its development.

rockets. He could put five pounds into orbit fairly soon and a heavier instrumented object soon thereafter—should an official project be set up.

At an August 3 follow-up to the June meeting, the same Army-Navy team met at the Redstone Arsenal. Von Braun prepared a paper for the group titled "The Minimum Satellite Vehicle Based upon Components Available from Missile Development of the Army Ordnance Corps." The ONR began scientific studies. Overall leadership was assigned to George Hoover, commander of ONR.

The satellite was nicknamed Project Slug, later to be called Orbiter. Envisioned was a five-pound satellite in Earth orbit; the Slug's advantage over its competitors was that it made use of existing technology and proven hardware and could go into space quickly.

Slug involved the Rocket Team, in cooperation with the Caltech's Jet Propulsion Laboratory (JPL) and the ONR. The price tag on the launch would be, by von Braun's reckoning, about $100,000. The Orbiter satellite would be carried by a Redstone missile as its first stage, a cluster of twenty-four or thirty-two Loki rockets as its second stage, a cluster of six Loki rockets as its third stage, and a single Loki rocket, attached to the satellite payload, as its fourth stage.

The disadvantage was that a five-pound satellite would not do anything and would be very hard to see. As Smithsonian historian Michael J. Neufeld later noted, "Von Braun and company apparently assumed that a 5 lb. minimal satellite was too small to carry even a radio beacon, a decision that will later prove damaging."

Over the next few months, as the International Geophysical Year took shape, America was preparing to make a commitment to launch a satellite. However, the inert Slug hardly fit the criteria of the IGY ruling body regarding "the launching of small satellite vehicles, to their specific instrumentation and to the new problems associated with satellite experiments, such as power supply, telemetering and orientation."

At the end of 1954, the public was getting a totally mixed message. On November 17, Secretary of Defense Charles E. Wilson denied knowledge of any American scientists working on Earth satellites and stated that he would not be alarmed if the Soviets built one first. The Department of Defense, in a two-sentence comment issued on December 21, reported that studies continued to be made in the Earth satellite program as the Orbiter group worked on a detailed production and launch schedule for their first series of satellites. The schedule included an initial series of four satellites, the first of which was forecast to be launched in September 1956 at the earliest.

Selecting the Satellite Program

Historian R. Cargill Hall has called the first nine months of 1955 "perhaps the most momentous period for the inchoate American space program." On Valentine's Day, a few months before the decision was made to participate in the IGY satellite program, the Technological Capabilities Panel proposed launching a scientific satellite to establish the principle of "freedom of space." The panel was headed by James R. Killian Jr. of MIT. Its members had been assigned the task of looking into the chances of a surprise nuclear attack; they concluded that reconnaissance satellites would prevent such an attack and that a scientific satellite would establish the legal precedent of overflight.

With the endorsement of President Eisenhower and Secretary of Defense Wilson, the U.S. National Committee for the International Geophysical Year approved the idea of launching scientific satellites during the IGY. Selection of a suitable satellite program was assigned to the assistant secretary of defense for research and development, Donald Quarles, a man known for his brilliant understanding of both technical and political matters. This decision was not made public.

A month later, von Braun and 102 of his colleagues and their families received American citizenship at a mass ceremony in Huntsville—this meant that von Braun and company could now be fully cleared to work with secret and top-secret material for the satellite program without any restrictions.

Quarles received three separate proposals for the launching of the first U.S. satellites in July. He formed an advisory group to settle the issue. This eight-member committee was made up of two Quarles selectees, two Army delegates, and two electees each from the Navy and Air Force. Homer J. Stewart of JPL was appointed head of what became known as the Stewart Committee. The Stewart Committee conducted an exhaustive analysis of the three plans submitted: the Army-led Orbiter; a late-entry Air Force modification of its Atlas missile to carry satellites; and a proposal from the Naval Research Laboratory (NRL) to modify a Viking research rocket for satellites.

The NRL's proposal was the brainchild of Milton Rosen, chief engineer of the Viking sounding rocket program. Rosen had come up with a plan for the Vanguard satellite with a scientific package that would satisfy the needs of the scientific community and the IGY. The Vanguard could be launched by a much smaller rocket based on Viking's technology and efficiency.

On August 24, the Stewart Committee selected Project Vanguard as

America's sole IGY satellite program. The vote was 5 to 2, with one member absent. Stewart himself voted in favor of Project Orbiter.

Many factors were at play in the decision to select Vanguard over Orbiter. One factor was a 1955 directive—issued by the National Security Council and approved by President Eisenhower—which decreed that the U.S. program could not employ a launch vehicle currently intended for military purposes. This decision was made in part to underscore American peaceful intentions in space, but also to keep ballistic missile programs from being diverted to or disrupted by the satellite program.

Proponents of Vanguard claimed that their program would not interfere with ongoing ballistic missile research. They also pointed to the fact that the Viking research rocket had already proven its worth as a scientific carrier vehicle. Orbiter was rejected primarily because the program was considered to offer a poor tracking system and inferior scientific value. In addition, Orbiter would require Redstone and Sergeant military rockets, violating the desire of President Eisenhower that U.S. scientific satellite research be distanced from the military.

Publicly, Eisenhower had made it clear that he wanted to keep the military out of the IGY program, which was dedicated to scientific purposes. The Army's Redstone had a lethal image—when it would make its public debut the following February, it would attract the nickname "city wrecker" ("Army Unveils 'City-Wrecker' " was the headline in the *Cincinnati Post*). One writer called it "a weapon that could carry atomic death and destruction to an enemy target 1500 miles away." Eisenhower feared that the rest of the world would focus on an Army missile being used to launch a scientific IGY satellite, and he had no interest in a satellite that could easily fuel anti-American propaganda. Eisenhower believed that a satellite put up as part of the IGY program would strengthen the concept of freedom of the skies and would be less likely to disturb Nikita Khrushchev if the United States kept the military out of the IGY program.

The Army, Navy, and Air Force were formally notified on September 9 that the Stewart Committee had approved the NRL satellite program and that the 5–2 decision had been reviewed and approved by the Department of Defense. Project Orbiter was killed, and the Army was forbidden from attempting the launching of satellites.

Four days later, a different committee recommended to Eisenhower that the Army and Navy cooperate in the development of the Jupiter intermediate-range ballistic missile (IRBM) to be deployed by the Army on land and the Navy at sea. The Army was given a chance to develop this four-stage version of the Redstone—not to launch satellites but to support tests of the Jupiter nose cone during high-speed reentry tests. Called Jupiter C for "Jupiter Composite Re-Entry Test Vehicle," the first Jupiter C was expected to be launched by September 1956. This group was headed by MIT's James R. Killian Jr., an adviser to President Eisenhower on science and technology.

The project was small consolation for von Braun, but it kept the Army Rocket Team in the running. What von Braun and later his commanding officer, Major General Medaris, had failed to realize from the outset was that the Pentagon had little interest in the satellite project, especially von Braun's. Even with White House backing, Vanguard got short shrift.

As legendary aviator Charles A. Lindbergh recalled: "In 1955, I was sitting as a member of one of the ballistic-missile scientific committees when the subject of orbiting a satellite was broached. Everyone present agreed that a satellite-launching project was practical, important, and deserving of support. But what priority should be assigned to it? We were working under staggering concepts of nuclear warfare, and the imperative need to prevent a strike against the United States by the Soviet Union. Our mission was to speed the development of intercontinental ballistic missiles with sufficient retaliatory power to discourage attack. Members of the committee believed that the security of our nation and civilization would depend on our ability to shoot across oceans quickly and accurately by the time Russia had missiles available for shooting at us. A shortage of scientists, engineers, and facilities existed in fields of missiles and space. Military projects needed every man available. The consensus of committee opinion was that we should concentrate on security requirements and assign to the Vanguard program a secondary place. There would be time to orbit satellites after our nuclear-warhead missiles were perfected and adequate marksmanship achieved."

The Vanguard contract was awarded to the Glenn L. Martin Company in Middle River, Maryland, to develop a three-stage, nonmilitary rocket specially for the IGY satellite program. The first-stage booster was an improved version of the Viking research rocket that had been used for the first time in 1949. Fifteen vehicles were scheduled for the program—two Viking rockets and thirteen of the improved Vanguard version would be able to develop 30,000 pounds of thrust from liquid fuel. Before the rocket exhausted its fuel supply at an altitude of about 36 miles, it would attain a velocity of approximately 3,700 miles per hour. The second stage would then take over and carry the satellite to an altitude of 300 miles at a speed of 9,000 miles per hour, at which point the solid-fuel third stage would kick in, pushing the satellite into orbit at 18,000 miles per hour.

Vanguard was placed under Navy management and Pentagon monitorship. The program's objectives were to develop and procure a satellite-launching vehicle; to place at least one satellite in orbit around the Earth during IGY; to accomplish one scientific experiment; and to track the flight of the satellite.

Late in 1955, after a statement came from the Pentagon officially announcing that the Navy was to take charge of the satellite program, the Navy publicity machine began telling the story of how it would be humanity's agent for entering space. Project Vanguard was strictly a civilian program, and, according to John P. Hagen, the civilian astronomer in charge, the rules were simple: Vanguard could not benefit from either existing military missile research or military technology. There was to be no secrecy involved—in fact, every development was to be recorded and published (although engineering reports that were candid about Vanguard problems were kept officially confidential until 1963).

From the outset, the mission plan was direct enough to be set forth in an official statement consisting of a mere thirty-two words: "1. To place a scientific Satellite in an orbit around the earth. 2. To prove that the Satellite is in the orbit. 3. To conduct a scientific experiment in the upper atmosphere using a Satellite." That was it. No more, no less.

The Vanguard project moved ahead—slowly.

Jupiter Rising

The Army Ballistic Missile Agency (ABMA) was established at Redstone Arsenal at the beginning of February 1956 under the command of Major General John B. Medaris. The unit inherited Wernher von Braun's Rocket Team and the projects already under way at Huntsville, including the Redstone missile.

Two months later, von Braun, with Medaris as an eager and determined partner, again indicated interest in the satellite-orbiting program by announcing that the ABMA's new Jupiter C missile, adapted from the Redstone, could be used to place a small Orbiter satellite in orbit by January 1957—ahead of Vanguard and before the beginning of the IGY. However, the assistant secretary of defense informed Huntsville that it was not to initiate any "plans or preparations" for using any part of the Redstone or new Jupiter programs as the basis for an orbital launch vehicle. As Vanguard developed, the Huntsville group under Medaris and von Braun was turned down repeatedly. In despera-

Major General John B. Medaris holding a model of the Redstone rocket at a press conference at the Fairmont Hotel in San Francisco on May 17, 1957. (U.S. Army, courtesy of Medaris Collection, Florida Institute of Technology)

tion, they requested that Jupiter C at least be designated the backup to Vanguard. But a special assistant to the secretary of defense vetoed even this request, explaining that the need for ballistic missiles for retaliatory strikes was a national priority and that trying to meet two or more projects simultaneously might dilute the Army's tactical and intermediate ballistic missile work. It was a stunning defeat for von Braun, but it did not stop him.

The Army Rocket Team was anything but timid. On August 15, it demanded still another hearing. In addition to asserting that Project Orbiter could be revised to address all the weaknesses perceived by the majority of the Stewart Committee, it issued a memorandum critical of Vanguard, citing its low probability for success and the overly lengthy time frame for developing the launch vehicle. The Vanguard group was stunned and quickly rebutted this memorandum.

On September 20, 1956, the first Jupiter C was launched from Cape Canaveral with its fourth stage loaded with sand. The Jupiter C successfully reached an altitude of 682 miles, a speed of 13,000 miles per hour (Mach

18), and a distance of 3,355 miles—enough to get the fourth stage into orbit. However, because of the directive by the Department of Defense, the Jupiter C carried a dummy payload as its fourth stage with an engine filled with sand ballast instead of solid fuel in order to prevent the "accidental" boost of the fourth stage into orbit.

If the Jupiter C fourth stage had been allowed to go into orbit, it would have become an Earth-circling satellite prior to the officially sanctioned Project Vanguard and, of course, before Sputnik.

Von Braun and Medaris later maintained that Washington had been so fearful that ABMA would use this launch to get into space that two observers were sent from the White House Bureau of the Budget to ensure that von Braun and his team did not sneak something into orbit. "They had somebody in the control room from Washington so the fourth stage would not be ignited," said Bob Swinghamer to historian Mike Wright.*

As news of the incredible September launch slowly became public, rumors begin to fly. With no evidence to go on—other than that the Army supposedly had the wherewithal to have "accidentally" put something into orbit—it was suggested that the Army had actually put "something up." This was not as far-fetched as it might seem because, according to Sir Arthur C. Clarke, "on one occasion von Braun was preparing to launch a slightly clandestine satellite with a Jupiter-C (virtually a duplicate of the vehicle that eventually orbited the first United States satellite, Explorer 1), but the Department of Defense discovered the project in time to frustrate it."

On October 15 *Time* magazine carried an article on the progress of Vanguard to which was appended a paragraph that read in part: "Secrecy always breeds rumors, and a widespread rumor in the missile business was that the Army hoped to toss a satellite into the sky ahead of Project Vanguard which was administered by the Navy. Leader of this dark plot, according to rumor, was famed Wernher von Braun, chief creator of the V-2." Von Braun was said to think that the satellite-launching vehicle should have a more powerful first-stage rocket than the Vanguard, and he had such rockets, notably

*This comment appeared in the spring 1998 issue of *Alabama Heritage*. At the time, Swinghamer was associate director, technical at the Marshall Spaceflight Center in Huntsville. Others have asserted that Eisenhower himself was involved, but several historians I spoke with question whether Eisenhower actually put people in the control room. Real or imagined, this visit by Eisenhower's henchmen would come up later in Senate testimony to underscore the president's desire to let someone else be first in space. True or not, it was believed by those in Huntsville.

the mighty Redstone. "Unless stopped by higher authority, Army missile men may try to beat the Navy into space," the *Time* article concluded. Not only did ABMA not deny the story, but Medaris gave the clipping an honored place in his scrapbook. The article had all the earmarks of a leak by Medaris, who was frustrated over not being able to tell the world about Huntsville's September triumph with a missile able to fly over 3,000 miles.

On November 26, 1956, the Army Ballistic Missile Agency suffered a stunning blow. Secretary of Defense Wilson created specific missile roles for each of the three services. The Navy was to have control over ship-based missiles, the Air Force would have jurisdiction over long-range missiles, and the Army would be limited to missiles traveling less than 200 miles. There was no mention of space. Jupiter deployment was assigned to the Air Force, with ABMA acting as contract supplier only. "The psychological impact at Huntsville was devastating, not only at the Arsenal but in the town," Medaris later recalled, "and morale reached an all-time low."

Vanguard had been selected at a moment when the services were locked in a long-standing, debilitating, and unprecedented rivalry. To some of those involved—including the old Army man in the White House—the military myopia was so intense that the interests of a particular branch of service outweighed the common good.

The counterproductive rivalry was evident in late 1956 when a thirty-five-year old Marine Corps pilot named John Glenn was working on a new frontline jet fighter for the Navy called the F8U Crusader. In order to prove the value of the new plane, he decided to try and establish a new speed record flying across the United States. This quest would be known as Project Bullet.*

Project Bullet's 2,445-mile flight was set for July 16, 1957, and would require three midair refuelings. The Navy had tankers to refuel jets, but they were slow, propeller-driven planes. The Air Force had tankers that could fly higher and faster and would save about twenty minutes on the transcontinental flight. Glenn requested the help of the Air Force, which responded: "We

*In his 1999 memoir, Glenn recalled how he came up with the name: "When I realized that the Crusader flew faster than the muzzle velocity of a bullet shot from a .45 caliber pistol—586 miles an hour—I had a name: Project Bullet." Glenn was given a hero's welcome in New York and much press coverage, including a major profile in the *New York Times*. He also got to appear on the CBS quiz show *Name That Tune*. The newspaper noted, "At 36, Major Glenn is reaching the practical age limit for piloting complicated pieces of machinery through the air."

have carefully considered your request and would normally welcome the opportunity to take part in Project Bullet. However, our existing operational commitments, now and in the near future, are such that participation was impossible."

As Glenn later recalled, "Reading between the lines, it was fine with the Air Force if Project Bullet failed and the Air Force kept the record." But the Air Force did not keep the record, and Glenn landed at Floyd Bennett Field on Long Island in three hours, twenty-four minutes, and 8.4 seconds, breaking the record by twenty-one minutes. Even slowing down to refuel at 300 miles per hour (mph), Glenn averaged 723 mph, which was 63 mph faster than the speed of sound at 35,000 feet and considerably faster than a bullet.*

As it turned out, Wilson's decision to give the branches of the armed services very specific roles and missions with regard to missiles exacerbated rather than assuaged the rivalry. To old Washington hands, this was a replay of the immediate post–World War II military battle for control of the atomic bomb, which finally had to be given to a civilian agency, the Atomic Energy Commission, created for the purpose of impartial stewardship of all things nuclear.

In the winter and spring of 1957, the Rocket Team and ABMA faced a terrible dilemma. Its missile role had been trivialized, and its eagerness to go into space was thwarted at every turn. It would continue to push for its place in both realms.†

Huntsville continued to test and develop a missile that the Army could not use because its range exceeded the 200-mile limit imposed by the secretary of defense. Despite this, the Jupiter C was turned into a show horse. On August 8 a Jupiter C booster, launched by the Army at Cape Canaveral, at-

*The sonic booms created with the help of perfect atmospheric conditions extended over the eastern third of the United States. As he recalls with some glee in *John Glenn: A Memoir*: "A ceiling collapsed in Indiana, windows broke in Pittsburgh, and people grumbled. Somebody called the Pennsylvania National Guard about an unexplained explosion."

†But von Braun still would not let go of his plan. Everyone who came within hailing distance of the charismatic German got the pitch—even after the first Sputnik went up. Charles A. Lindbergh later recalled: "During my visit to the Army's Redstone Arsenal at Huntsville on a military mission, Wernher von Braun showed me one of his mock-ups and told me of his plans for orbiting a satellite. He had been unable to obtain the authorizations and funds required to put an actual satellite in orbit; but according to his estimates, he could have done so many months before Russia's Sputnik I was launched—at a minute cost as measured by present-day space expenditures."

Hermann Oberth in the foreground with officials of the Army Ballistic Missile Agency at Huntsville, Alabama, in 1956. Behind Oberth are Dr. Ernst Stuhlinger (seated), Major General H. N. Toftoy, commanding officer and author of "Project Paperclip," Dr. Eberhard Rees, deputy director, and Wernher von Braun, director. (NASA)

tained an altitude of 600 miles and a range of over 1,300 miles. Recovery of the scaled-down nose cone marked the first recovery of an object from outer space. It also marked a successful Jupiter nose-cone test, proving that ablative-type nose cones could survive the intense heat of reentry.*

Finally, on August 21, 1957, Major General Medaris ordered that remaining Jupiter C components be placed in protective storage. (Just three of the planned twelve Jupiter C rockets had been launched to date.) This was a move aimed directly at providing an as-yet-unauthorized Army backup to Project Vanguard. Medaris ordered that the two Jupiter C rockets in the most advanced state of readiness be put on standby in case a satellite launch was approved. The rockets would be maintained so that the first could be launched not more than four months after authorization to launch a satellite was given. The second could be launched about a month later. ABMA aspirations to launch satellites with these Jupiter C components were, for the time being, kept secret. At the same time, scientists at the Jet Propulsion Laboratory secretly had been working on the satellite itself.

*This nose cone would be displayed on television from the Oval Office by President Eisenhower on November 7, more than a month after Sputnik, to show the nation it had solved the problem of reentry from space.

Von Braun waited in the wings—bitter, but ready to go. He had held the Rocket Team in place and added its patron saint to the roster. Hermann Oberth, the inspiration for and the original member of the Rocket Team and the last surviving member of the trinity of pioneers in astronautics, left his home in West Germany and joined his colleagues in Huntsville with little fanfare in 1955. Von Braun had gotten permission to hire him after it became known that he had turned down an offer from the Russians who had offered him the chance to move to East Germany and do research there. Oberth's position in America was less than ideal. "Once in Huntsville, Oberth found himself in an environment that would have appealed to the imagination of Franz Kafka," write Frederic I. Ordway III and Mitchell R. Sharpe in *The Rocket Team*. "He went to work for the Army performing advanced space studies. The results of his work were often classified secret. Since he was not a citizen of the United States, he could not be granted security clearance for military secrets. Thus, he was unable on occasion to have access to his own work, and so on ad infinitum."*

The public paid scant attention to all of this. The fact was that in polls conducted in 1957 most Americans did not know what a satellite was. The *Baltimore Sun* determined that at the time of the Sputnik launch only about 15 percent of the local population was aware that the Martin Company was building the Vanguard. Marylanders and their fellow Americans were hardly alone in their ignorance of things orbital. Two years before Sputnik, the British astronomer royal, Richard Wolley, had dismissed the concept of spacc travel as "utter bilge."

*In 1958, Oberth returned to Germany to collect his old-age pension, which he would have forfeited had he not returned.

CHAPTER FIVE

The Birth of Sputnik

> For reporters who thought they knew the score, history's lens can be eye-opening. That's what newly declassified secret documents have done for some of us who covered the dawn of the space age. It seems that neither savvy science writers, the scientists who advised them, nor the leadership of the Soviet Union knew that, when Sputnik 1 caught most Americans napping on Oct. 4. 1957, it also helped fulfill one of the Eisenhower administration's secret strategic goals. It helped legalize the future use of spy satellites.
>
> —Veteran space reporter Robert C. Cowen,
> *Christian Science Monitor* of October 13, 1996

On May 26, 1954, Sergei P. Korolev, chief designer of the Special Design Bureau, first proposed to the leadership in the Kremlin that the Soviet Union launch an artificial satellite of the Earth using the R-7 rocket. The new rocket had been approved a year earlier and was being developed to carry a warhead weighing five tons over an intercontinental ballistic trajectory. Planned to be four times more powerful than von Braun's Redstone, the R-7 could potentially place a satellite weighing 1.5 tons into orbit. To avoid several rocket stages, the Soviets opted to go with a cluster of rockets around a central core. These clusters would be ejected after they had used up their fuel, while the central core motor continued to burn.

On August 30, 1955, one month after the United States and the USSR both promised IGY satellites, Korolev got the green light to launch a satellite.

The USSR's workhorse R-7 rocket, employed from Sputnik I and adapted as the launch vehicle for other important programs. Depicted is a launch of July 15, 1975, carrying a crew of cosmonauts who would link up with Apollo astronauts in Earth orbit as part of the Apollo-Soyuz mission. (NASA, courtesy of the USSR Academy of Sciences)

At the beginning of 1956 the Council of Ministers approved a program leading to the launch of Korolev's first artificial Earth satellite early in the International Geophysical Year. By late September 1956 Korolev had designed his artificial satellite—a heavily instrumented 3,000-pound geophysical observatory that would be a stunning contribution to the IGY. Work on it began.

However, in January 1957 Korolev had a change of heart. With an eye on news reports coming out of America on Vanguard progress, he suggested that he be allowed to launch two small satellites before the year was over, ahead of the large one he originally proposed. Quite simply, he wanted to beat the United States into space. His accelerated plan was quickly approved; the satellite that originally was to have been Sputnik I was now third in line and scheduled to orbit in early 1958.

The first R-7 exploded on the launchpad in the spring of 1957, followed by five more failures. On May 15, 1957, for example, an R-7 rocket was launched from the Baikonur complex (not far from Tyuratam, near the Aral Sea in the Kazakh Republic). Its ballistic trajectory was aimed toward the Pacific Ocean, but one of the strap-on boosters detached ninety-eight seconds into the mission and the rocket had to be destroyed. When the IGY began in July, the Soviets still had no working R-7. Just over a week later, another R-7 disintegrated—this time a mere thirty-three seconds into the mission.

In August the R-7, in its sixth attempt, was launched from the Baikonur complex into a ballistic trajectory toward the Pacific Ocean. The mission was a success, although its dummy nuclear warhead broke up as it reentered the atmosphere near its intended destination of the Kamchatka peninsula.*

With this success, an eager Korolev moved quickly. Despite some resistance, he was finally given permission to put the first Earth satellite in orbit. He hoped to be able to launch Sputnik in time to honor the centennial of the birth of his idol, Konstantin Tsiolkovsky, the father of Soviet space exploration, on September 17.

Russian Space Plans Are an Open Secret

Five days after the R-7's success, the USSR reported to the world that it had fired accurately an intercontinental ballistic missile. The text of the Soviet announcement read in part: "A super-long-distance intercontinental multi-stage ballistic rocket was launched a few days ago. The tests of the rocket were successful. They fully confirmed the correctness of the calculations and the selected design. The rocket flew at a very high, unprecedented altitude. Covering a huge distance in a brief time the rocket landed in the target area. The results obtained show that it was possible to direct rockets into any part of the world."

The announcement was not taken seriously in the United States. People were downright dismissive, perhaps because the Russian propaganda machine had been known to stretch the truth. For many years, the Russians claimed, for example, that they had invented baseball and took credit for other Western discoveries as well.†

If the U.S. government and the public had taken the Soviets seriously, they would have noticed that the Soviets had announced quite clearly years earlier their intention to place a satellite in orbit and had even placed in American hands the means to track the satellite when it went up. Between 1951 and October 1957, there were more than twenty occasions on which a

*James E. Oberg pointed out in his book *Red Star in Orbit* that Kamchatka was not too far from the Kolyma gold mines, where Korolev had worked as a slave laborer in 1938–39.

†As late as July 20, 1987, the *New York Times* reported on an article in *Izvestia* in which writer Sergei Shachin, citing cultural historians, insisted that baseball came from an ancient bat-and-ball game called *lapta,* which had been brought to what was to become California by Russian émigrés in the late eighteenth century.

Russian or someone talking for the Russians issued a public statement that they were headed into space. No fewer than twenty-five articles in the *New York Times* during this period discussed, in whole or in part, Russian space plans. The first *Times* mention came on October 4, 1951—six years to the day before Sputnik—when Soviet rocket engineer Mikhail Tikhonravov announced that Soviet technology was at least on a par with that of the United States and could well result in the launching of satellites in the not too distant future.*

During a speech in 1953 at the World Peace Council in Vienna, A. N. Nesmeyanov of the Soviet Academy of Sciences stated that "science has reached such a stage that the launching of a stratoplane to the Moon, the creation of an artificial satellite of the Earth, is a real possibility." On March 24, 1954, the *New York Times* reported on a Moscow radio broadcast urging Soviet youth to be first to reach the Moon. On April 15, 1955, an official Soviet government announcement confirmed that the Soviet Academy of Sciences had established the Permanent Interdepartmental Commission for Interplanetary Communications, with the short-term goal of launching instrumented weather satellites sometime over the next two years. In June of that year, Eric Bergaust, editor of *Aero Digest*, warned that the Soviet Union was a lot closer to space conquest than was the United States.

In June 1957, as the date of the Sputnik launch drew closer, the Soviets held press conferences to let the world know of their intentions. One document sent to an American International Geophysical Year official by an official of the Soviet Academy of Sciences not only spelled out such details as the expected orbiting speed of the satellite and its approximate launch site but predicted that the first one would fly within a few months. During the same month, the *New York Times* carried seven reports of Russian plans to go into orbit; one of them said that early satellites would carry dogs as passengers, and another claimed the first satellite would be about eighteen inches in diameter and weigh 100 pounds. The *Times* wasn't the only U.S. newspaper writing about Soviet satellites. "Russ Report All Ready for Satellite Launching" was a *Los Angeles Times* headline of June 2, and the *Washington Post* carried a June 10 prediction that the satellite would circle the world every ninety minutes and would be visible "like a star, through ordinary field glasses."

Also that month, Lloyd V. Berkner of the U.S. National Committee for the IGY was sent a seven-page, single-spaced memo signed by "Academi-

*On October 12 of the same year, the *Washington Post* noted that a report from Russia was predicting a Soviet trip to the Moon in ten to fifteen years.

cian I.P. Bardin, Vice-President, USSR Academy of Sciences, President, Soviet IGY Committee." It included a detailed discussion of what the Soviets hoped to achieve in both sounding rockets and man-made satellites during the IGY. They made no bones about their prowess in the field, as the memo noted: "Vertical sounding of the atmosphere by means of rockets has been in progress in our country for a number of years." In fact, a month earlier one of these rockets had carried a payload of 4,850 pounds to a height of 132 miles. The payload included a capsule containing two dogs, who were parachuted safely back to Earth. The memo did not say when the first satellite would go up. The firing of the rockets and the satellites would be conducted "approximately evenly" throughout the IGY. Since the IGY began in July, it was easily inferred that the first Soviet satellite would go up in a few months—say, September or October.

At about this time, A. A. Mikhailov, chairman of the Astronomical Council of the Soviet Academy of Sciences, wrote an article in the *Astronomical Journal, USSR* alerting astronomers that the first satellite would "have a visible stellar value from 4 to 9 and [a] period of rotation of about one hour and a half." By giving it an estimated visibility value of between 4 and 9 the Academy was hedging its bets on whether it could be seen by the naked eye—objects up to the sixth value can be seen without the aid of a telescope. Astronomers in the Soviet Union were requested "to be prepared for participation in the visual observations of the artificial satellites."

At the beginning of July, Soviet announcements and predictions mysteriously stopped. They picked up again in mid-September when Radio Moscow reported that the Soviet Union was now ready to launch a satellite. No launch date was specified, but the satellite would be heavier than the one planned by the United States.

In part, U.S. space officials and members of the scientific community discounted the Soviet claims because of their vagueness. "Basically, in the meetings of the international IGY committees, the Soviets kept refusing to give any real data," recalled mathematician and NASA official Homer E. Newell. "How could you interpret that? Well, either they didn't want to cooperate, or they weren't ready to make a launch. Everybody interpreted it as the latter."

But it could be argued that the Russians were more forthcoming than the Americans were able or willing to recognize. On August 27, academician V. A. Kotelnikov appeared at a conference in Boulder, Colorado, and would give no dates for the launching of the Soviet satellites but said that they would cross higher latitudes than the proposed American satellites and would broadcast at an output power of one watt on frequencies of about twenty and forty megacycles.

The noted British physicist-astronomer Sir Bernard Lovell later attributed the national shock that followed Sputnik's launch to "American blind disbelief in the powerful advance of Soviet science and technology."

To a large degree, Congress was especially reluctant. Prior to the launching of Sputnik, Lieutenant General James M. Gavin, who was then part of the Army's Weapons Systems Evaluation Group, had a hard time convincing Congress that launching a satellite was possible either by the United States or by the USSR. Finally, he decided to bring von Braun with him before a Senate committee. During his testimony von Braun began describing the Soviet space capabilities. Suddenly, Senator Allen Ellender of Louisiana broke in to ask if von Braun and Gavin were out of their minds. Ellender had just come back from the Soviet Union, and after seeing the decrepit, ancient automobiles—and very few of them—on the streets, he was convinced that the two men were totally wrong.*

Gavin was one of the few who actually tried to sound the alarm, but his warnings were not taken seriously, including a portion of his memo of August 15, 1955, sent to the assistant secretary of defense for research warning of the "risks of psychological damage if the Russians were first to launch." Other American public voices warning of the Russian rocket and satellite advances could be counted on one hand. "These Soviet developments should not be underestimated," wrote George P. Sutton, of the Aerophysics Department of North American Aviation, in an article in the January 1, 1954, issue of *Automotive Industries* after reporting that the Russians were working on "high-thrust, large rockets for long-range missiles or satellite vehicles."

Paving the Way for U.S. Spy Satellites

However, at the nonpublic level, the U.S. government was primed and prepared for Sputnik. RAND Corporation studies of unclassified Soviet literature led to an educated guess in mid-1957 as to when a satellite would be launched. The estimate was off by only two weeks. Central Intelligence Agency director Allen Dulles recalled in *The Craft of Intelligence* that "despite the general impression to the contrary, the intelligence

*Ellender would later be the only member of the Senate to vote against the bill creating the National Aeronautics and Space Administration.

community predicted with great accuracy Soviet progress in space technology and the approximate time when their satellite would be orbited."

Richard M. Bissell Jr., master of the U-2 program and the man in charge of airborne reconnaissance for the CIA, recounted in his memoirs: "In June 1957, three months before the launch of Sputnik, a U-2 pilot flying over Kazakhstan spotted something in the distance that caught his attention. Departing from his course, he stumbled on the crown jewel of Soviet space technology, whose existence had not even been suspected. Within a week of the U-2's return, a team of intelligence analysts, photo interpreters, and photogrammetrists had built a scale model of the Tyuratam test site. Photo intelligence also allowed analysts to determine the size and power of Soviet rockets based on burn marks and the configurations of the pads for exhaust gases." Ray S. Cline, later to become deputy director of the CIA, briefed President Eisenhower that the Russians were likely to send a series of small satellites into space.

Because of its stake in satellite reconnaissance, the CIA took an interest in Vanguard. By late 1956 Bissell realized that the American effort to get into space was lagging behind that of the Soviet Union. Bissell went to CIA chief Dulles and urged him to take action. Dulles sent him to see Assistant Secretary of Defense Donald Quarles to inform him that the CIA was interested in accelerating the development of a satellite. Quarles turned down CIA money but gave some extra funds to the Naval Research Laboratory. Bissell later speculated that Quarles might have felt "threatened" by his approach, but it was more likely that the assistant secretary believed that CIA money could compromise the scientific mission of the satellite by suggesting that it was being used to spy. When in April 1957 Vanguard experienced cost overruns, the National Science Foundation and the CIA funneled money into the program ($5.8 million and $2.5 million, respectively).

Why, then, was the CIA involved in Vanguard and why didn't the agency spend its money to support von Braun and Orbiter? Several historians have surmised that the CIA was helping to fulfill one of Eisenhower's strategic goals—the one that he had failed to achieve with Open Skies. Walter A. McDougall (in *The Heavens and the Earth*) and R. Cargill Hall (in several articles; see bibliography) reached this conclusion after examining primarily unclassified documents. In "A Strategy for Space," published in the British journal *Spaceflight* (September 1996), Dwayne A. Day of the Space Policy Institute of George Washington University used several dozen newly declassified documents to prove "that being first to launch a satellite was not only unimportant, but less important than achieving significant science." Eisenhower reasoned that a satellite put up as part of the IGY program would strengthen the freedom of the skies policy at a time when the United States was secretly deploying spy planes—the maiden U-2 Soviet overflight took place on July 4, 1956—and was developing spy satellites to reconnoiter the USSR.

It was crucial that the satellite be perceived as entirely scientific. A major reason von Braun's Rocket Team never had a chance was they were linked with a ballistic missile—if the United States used a missile to launch an inert, noninstrumented satellite before the beginning of the IGY, it would have been seen as a preemptive strike. But if the Russians beat the Americans and flew over the continental United States, they would, in fact, establish that space was an open platform and that national sovereignty did not extend beyond the atmosphere. Although there were those in the Pentagon and on the White House staff who desperately wanted to beat the Russians into space, the president was not one of them.

The Russians Extend an Olive Branch

The United States had been offered a chance to participate in the Russian IGY space program. Soon after the July 1, 1957, beginning of the IGY, the Russians approached the Naval Research Laboratory with an invitation to fly a radio frequency mass spectrometer on a Russian rocket. This was confirmed in an undated memo embedded with other dated memos from the days prior to the Sputnik launch. The memo was not disseminated beyond a few top officials. The original routing of the memo was to the head of Vanguard and three other top officials. For forty years the unclassified memo was part of a classified "Sputnik" file, which was inherited by NASA when it took over the rocket arm of the Naval Research Laboratory. The file has been declassified.

The memo from the chief of navy research, Robert M. Page, to the assistant secretary of the Navy for air (later to be renamed "for research and development") said in part: "A most careful review of many factors has led to the conclusion that it would be unwise to undertake this project at this time. The principal reason for this conclusion is technical."

Page then pointed out that it was doubtful the requested instrument could be designed during the remainder of the IGY so it would withstand the anticipated but not specifically known environmental stresses and still perform satisfactorily. "It is believed that it would be particularly unfortunate to attempt the experiment in a Russian rocket and have it fail," wrote Page. "The design problem is made virtually insoluble by our lack of information on the technical characteristics of the Russian rocket."*

In July 1957 a paper titled "USSR Rocket and Earth Satellite Program for the IGY" was presented by a Professor Petroff to the Americans. It included an IGY shopping list of achievements that were to be attempted, including the study of radiowave passage through the ionosphere "with a view to the determination and verification of its structure." The instrument for this was the aforementioned mass spectrometer.

The memo did not clarify whether this was an invitation to fly on a satellite or on a suborbital sounding rocket. There was only mention of a rocket, rather than a satellite, suggesting a sounding rocket. What the Russians called sounding rockets were their geophysical rockets beyond the capacity of the German V-2 or America's Viking. But it was more likely that the invitation was to fly on the heavily instrumented Sputnik III, which would not go up for almost a year. If the instrument was not intended for Sputnik III, it was likely intended for a mission that flew on February 21, 1958, with a 3,350-pound payload of instruments to study the Earth's magnetic field, the nature of the atmosphere, radiation, micrometeoroids, and the ionosphere.†

Even if the invitation was for a sounding rocket, it would still have been a remarkable breakthrough. The invitation probably was made before the Russians had determined which mission would carry the mass spectrometer. One thing was almost certain, however: The offer came from Soviet scientists rather than propagandists, because the propagandists would have used it to demonstrate American uncooperativeness in the face of Soviet goodwill.

Regardless of whether it was a rocket or satellite offer, the invitation appears to have been made in the best spirit of the IGY and belied many of the later claims that a bellicose Russia had turned the IGY into a display of military strength. The Russian overture was in keeping with two official Rus-

*The major reason that there were lines of communication in place for scientists of both nations was that they were cooperating in Antarctic research. In 1956 Soviet scientists had "wintered over" in the American base, Little America, while the Americans lived at the Soviet post Mirnyy.

†Although suborbital sounding rockets are called the "drudges of rocketry," their value had been immense. Their major flaw is that they have little of the glamor of carrier rockets, which are, as von Braun and Frederic I. Ordway III put it in their history of space travel, "the apotheosis of rocketry—the culmination of centuries of aspirations."

sian statements that appeared in the *New York Times* on consecutive days in 1955. On August 2 Premier Khrushchev reported that the USSR would cooperate with the United States if the project was in the "interest of man." The next day Leonid I. Sedov, the man who would become the spokesman for the Soviet space program, said that he envisioned cooperation with the United States in the Soviet satellite program.

In the months following the Sputnik launch, the voices that would be heard loudest in America would be strident ones, leaving little if any room for reflection on the point that Sputnik was an element in a larger international scientific project among scientists and other participants in the IGY. Ronald Potts, a consultant to the Naval Space Technology Program who spent part of the IGY working alongside Russian scientists in Antarctica, said in an interview in June 2000 that the scientific community's response to Sputnik and the IGY was totally at odds with the impression of politicians and journalists, who regarded it with alarm. "I cannot recognize many accounts of the period because they leave out the spirit of cooperation fostered by the IGY. [Page's] memo was consistent with my experience of the cooperation which was at the heart of the IGY."

After seeing the memo, veteran space and science writer Bill Hines said: "What the paper neglects to talk about—and it must have been an obvious factor in the decision it puts forward—was that should a mission come to pass, the Russians would have had total access to the U.S.-built payload during the prelaunch phase, which might well have given them some valuable information about the state of American communications technology." The objections framed in the memo were, to Hines, a cover for our unwillingness to share state-of-the-art technology. "The fact is we didn't trust the Commies then, and even at the time of Apollo-Soyuz, the element of trust involved in that mission was minimal. About the only 'secrets' the two sides gave each other in ASTP [Apollo-Soyuz Test Project] were engineering needed for design of the tunnel joining the two ships."

After turning down the Soviets' invitation, the United States proceeded with its own space plans, testing the first Vanguard payload on a sounding rocket on April 11, 1957. American U.S.-IGY scientific satellite equipment—including a radio transmitter and instruments for measuring temperature, pressure, cosmic rays, and meteoric dust encounters—was tested above Earth for the first time, as a rocket containing this equipment was fired by the Navy to a 126-mile altitude. On May Day there was a key suborbital test of the Vanguard Test Vehicle (TV-1). It carried a modified Martin Viking first stage and Vanguard solid-propellant third stage with a Grand Central rocket as a second stage, and was launched with an instrumented nose cone to an altitude of 121 miles. It met all test objectives. Finally, on October 1, 1957, Minitrack, the worldwide tracking system for the Vanguard, was declared to be fully operational.

America was preparing to go into space, but the Soviet Union was ready to go.

Countdown to the Launch

A month earlier, on Tsiolkovsky's centenary birthday, Sergei Korolev made a rare public appearance. Speaking at a rally at Pillared Hall in the Palace of Unions in Moscow, he announced that "in the very near future, the first experimental launchings of artificial earth satellites will take place in the USSR and the USA for scientific purposes!" The same day, the Soviet newspaper *Pravda* carried an article by Korolev on his hero. "Soviet scientists are working on problems of deep penetration into cosmic space," reported Korolev. "We are witnessing the realization of Tsiolkovsky's remarkable predictions of rocket flights and the possibility of flying into interplanetary space—predictions made more than sixty years ago."

After the rally, Korolev returned to anonymity at the Design Bureau and its special "Sputnik room," where work was being completed on the satellite. A few days later, he flew to Baikonur to direct preparatory work for the launching.

The Baikonur complex—later to be known as the Baikonur Cosmodrome—was, as one of the men stationed there called it, "a terrible piece of barren steppe," where summer temperatures reached 122 degrees Fahrenheit. Those stationed there lived in railway cars; there was so little natural water the men sometimes washed with bottled mineral water. Workers desperate for entertainment would catch scorpions, put them in glass jars, and watch them fight to the death.*

Although one of the early designs called for a cone-shaped satellite, it was, as historian James Harford puts it, "Korolev's aesthetic as well as engineering sense that had led him to insist on the ball shape for Sputnik 1." Ko-

*In America, that amusement had become distilled into a cautionary metaphor for the United States and the USSR. In 1953 nuclear scientist J. Robert Oppenheimer wrote, "We may be likened to two scorpions in a bottle, each capable of killing the other, but only at the risk of his own life."

rolev himself, in an interview published after his death in 1976, said: "It seems to me the first Sputnik must have a simple and expressive form, close to the shape of natural celestial bodies. It would forever remain in the consciousness of people as a symbol of the dawn of the space age."

Korolev's team worked around the clock and in little more than a month created and mounted a plain, highly polished, aluminum-alloy, 184.3-pound sphere containing only a radio transmitter, batteries, and instruments to transmit temperature and pressure readings from inside the satellite. It consisted of two hemispheric casings held together by a frame and was 22.8 inches in diameter. In the assembly hall, Korolev put the satellite on a special stand, draped in velvet, so that the workers would show reverence toward it. He personally supervised the carrying out of the production schedule every day. "Korolev came over to the shop and insisted that both halves of the Sputnik's metallic sphere be polished until they shone, that they be spotlessly clean," recalled Konstantin Feokistov, who would be the first engineer-cosmonaut to go into orbit in the three-man Voskhod I seven years later. In a conversation with James Harford, Feokistov said: "The people who developed the radio equipment were actually the ones demanding this. They were afraid of the system overheating, and they wanted the orbiting sphere to reflect as many rays of the Sun as possible."*

Korolev proceeded from the idea that a highly reflective ball would be easier to see with the naked eye. "It is important that the orbit and the optical characteristics of the satellite are such that almost all people in the world will be able to watch its flight with their own eyes," he later said. He was also obsessed with the idea that it should be polished to the highest level possible. He knew that he and his crew would be the last to see Sputnik because the satellite would burn up during its return to Earth.

The launching was originally set for October 6, 1957, but during the final preparations, Korolev decided to move it ahead two days to the fourth. He feared the United States might upstage the USSR at the last minute. His immediate concern was a paper (titled "Satellite in Orbit") being delivered in Washington at the IGY conference being held during the week beginning September 30. Despite assurances from Soviet intelligence to the

*Gennadi Strekalov, one of the metalworkers who were assigned to the Sputnik I manufacture, would also become a cosmonaut. "My teacher in metalworking did the finishing," he told James Harford. "Two half spheres were stamped, then machined, then the masters did the finishing." Strekalov, who in 1995 flew to the Mir space station with American astronaut Norman Thagard and then participated in the first Mir-shuttle rendezvous, was as proud of his work on Sputnik I as he was of his four orbital flights.

The first official photograph of Sputnik to be transmitted to the West, and the only known photo purporting to be of the actual satellite. (Author's collection)

contrary, he had a nagging fear that Vanguard was about to be launched and thought the paper had something to do with the launch.

Sputnik was placed inside a pointed metal nose cone and installed atop the gigantic R-7 rocket. On October 2, 1957, it was rolled out to the launchpad.

The actual moment of blastoff was delayed a number of times during the day, but finally, at 19:12 hours Greenwich Meridian time on October 4, from the Baikonur complex, the ground shook as no fewer than twenty rocket boosters thundered with 1,120,000 pounds of thrust. The one-and-a-half-staged rocket climbed almost vertically at first, slowly tilted down range, then accelerated to 17,400 miles per hour—an unheard-of speed.

Sputnik left the Earth housed in a protective nose cone in the forward section of the upper portion of the injection stage. After the upper stage reached an altitude of 142 miles and a velocity of nearly 26,240 feet per second flying almost parallel to the Earth's surface, the engine was shut down. The protective cone was jettisoned, and the satellite, separated from the orbital injection rocket, began its free flight in orbit around the Earth. The moment the satellite separated from the injection rocket, four pivot antennae, eight to nine and a half feet long, sprung loose to their working positions and two radio transmitters established an immediate link to the Earth below. Sputnik's orbit ranged from 141.7 miles at the closest point to Earth to 588 miles at the farthest.

Ninety-six minutes and seventeen seconds later, Sputnik passed to the east of its launchpad with its transmitter sending out a beeping noise that blared from Baikonur's loudspeakers. Cheers and shouts of joy exploded from observers on the launchpad. Korolev turned to his associates and said, "Today we have witnessed the realization of a dream nurtured by some of the finest men who ever lived, including our outstanding scientist Konstantin Eduardovich Tsiolkovsky." Later on, when Sputnik was installed in orbit, and its call sign was heard over the globe, he was reported to have exclaimed, "I've been waiting all my life for this day!"

CHAPTER SIX

Red Monday

Ignore the viruses of dread that float through family rooms: the hydrogen bomb erupting from the South Pacific like a cancerous jellyfish the size of God; or the evil Senator Joseph McCarthy and the evil Commies he never catches one of, not one, though he does manage to strew the land with damaged lives and the liberal tic of anti-anti-communism; or Sputnik, the first satellite, built by Russian slave labor, no doubt, while our top scientists were developing the Princess phone, 3-D movies and boomerang-shaped coffee tables.

—Pulitzer Prize–winning writer Henry Allen on the moods of the 1950s, *Washington Post*, September 27, 1999

On Monday October 7, 1957, America's mood turned sour. The public's surprise and awe over the weekend quickly changed to anger and shock as the impact of Sputnik was felt. The weekly newsmagazines helped set the tone: "Red Moon over the U.S.," *Time* said, editorializing over what it referred to as the "chilling beeps" of the satellite, while *Newsweek* noted that Moscow "had already given the word 'satellite' the implications of ruthless servitude," asking, "Could the crushers of Hungary be trusted with this new kind of satellite, whose implications no man could measure?"

Readers of the *Washington Post* woke to one Monday morning article titled "'Sputnik' Could Be Spy-in-the Sky," another proclaiming "Russia Plans

2nd Satellite . . . in 'Nearest Future,' " and a long and sobering editorial on the need for the United States to embark on the "conquest of space." Monday's *Washington Evening Star* suggested that this was a U.S. loss brought on by sluggishness and complacency; the *New York Herald-Tribune* called it nothing less than "a grave defeat."

On Monday night, Eric Sevareid began his CBS radio network broadcast: "Here in the capital, responsible men think and talk of little but the metal spheroid that now looms larger in the eye of the mind than the planet it circles around." Earlier in the day, *Washington Post* writer and analyst Chalmers Roberts summed up the three things on the minds of the authorities in Washington that Monday morning: that Sputnik would have an extreme impact on the leaders of the underdeveloped world, who would see it as a victory for socialism; that its surprising size and weight proved the Soviet Union had the power to launch and deliver an "intercontinental ballistic missile with a multi-megaton hydrogen bomb warhead of several thousand pounds" to any point on the face of the Earth; and that a big argument was about to break out in Washington as to what must be done and who was responsible.

Coincidentally, on October 7 disquieting news came from England that a fire in the Windscale plutonium plant north of Liverpool had spread radioactive iodine throughout the countryside. Although it would be some time before the consequences would be known fully (cancer would claim the lives of dozens living nearby), there was certainty that innocent people would die from radiation poisoning.

But of all that happened on that October Monday to sour the American mood, nothing did more to trigger hysteria than the Soviet testing of a newly designed "hydrogen weapon." As the news of this superweapon reached Washington the next day and spread through the nation, Sputnik seemed proof that the USSR was a disciplined society able to compete and beat the United States not only with the cutting edge of technology but with powerful weaponry.

Americans had underestimated Russia, confusing shoddy Russian cars and other consumer goods with the state of Soviet science and technology. Prior to Sputnik, when there was much talk of small transportable "suitcase bombs," a common joke was that the Russians could not surreptitiously introduce nuclear explosives into the United States because they had not yet been able to perfect the suitcase. After Sputnik, the laughter stopped.

An Abundance of Fear

In the days before the Sputnik launch, Americans outwardly seemed to be content. The U.S. population had passed 170 million in February as the "Baby Boom" continued at the lusty rate of 2.5 million births a year. The nightmare of the Great Depression was becoming a dim recollection. The long-dreamed-of minimum wage was now one dollar, and a gallon of gas was a mere twenty-three cents. Volkswagen sold more than 200,000 Beetles, and J. Paul Getty made one billion dollars. Americans were enthralled with television quiz shows featuring sweating contestants closeted in soundproof booths. Consumer goods flourished everywhere, and many Americans dreamed of eventually possessing a color television set. And there was more to American life than mere material goods: During a sixteen-week New York "crusade" ending in September, the Reverend Billy Graham had drawn an audience of 2 million souls.

Despite the outward appearance of calm and satisfaction, Sputnik appeared at a moment when America was anxious on several fronts. Sputnik did not occur in isolation, and the fear it engendered played on other fears. For starters, the bottom had fallen out of a seemingly indestructible economic boom. Federal spending and wages were going up, creating inflationary pressure and talk of recession. Stock prices, which had started to falter in the summer, had been dropping steadily in September.

Moreover, the crime rate for the first half of 1957—1,309,670 crimes, according to the FBI—was the highest on record. FBI director J. Edgar Hoover was suddenly talking about "emotional disturbance" as a precursor to the grimmer crime stories that were making it into the newspapers. In January, George Metesky, fifty-two, was arrested in Waterbury, Connecticut, after confessing to be the "Mad Bomber" who had planted thirty-two bombs that had injured sixteen people in the New York area. Prior to his arrest, more than 160 prank phone calls were made to the police from people claiming to be the "Mad Bomber." On New Year's Eve alone there were 30 calls. The bomb hoaxes spread to other U.S. cities.

Wisconsin authorities accused and convicted Ed Gein, a fifty-one-year-old bachelor farmer, handyman, and sometimes baby-sitter, of a series of horrific murders and grave-robberies. The obscene details of Gein's crimes were kept from readers of the more respectable newspapers and chronicled mostly in the tabloids. Gein was the inspiration for Buffalo Bill in *Silence of the Lambs*, Norman Bates in *Psycho*, and other fictional mass murderers.

Professional criminals were also making news. Albert Anastasia, the most feared man in organized crime, was shot on October 25, 1957, while sitting in a barber chair in New York's Park Sheraton Hotel. Reviewing the year's events, *The Unicorn Book of 1957* said: "A ruthless gang killer, a Mad Bomber and a psychopathic Bluebeard—together these three, who perhaps outdid crime fiction's most grisly characters, made the crime year 1957 gruesomely memorable." Topping it all off, in November police found fifty-eight mafiosi from all over the United States at a farm in Apalachin, New York, meeting like so many Rotarians.

Crime was hardly the nation's only source of anxiety. The 1957 influenza epidemic was abating by mid-October, but it had been the worst outbreak since the end of World War I, killing 70,000 Americans—mostly the young, the elderly, and the infirm. The normal flow of life was interrupted as picnics, college football games, and county fairs were canceled as a precaution against the flu.

The day Sputnik was launched was also the peak of the Cold War, barely six weeks after the Soviets had test-fired the first intercontinental ballistic missile, and less than a year after Soviet leader Nikita Khrushchev had made his "we will bury you" threat against the United States. On top of all this, and before Sputnik, the Russians had said they had the ICBM and Khrushchev declared that they were turning out nuclear missiles "like sausages." Security fears were such that on October 1, the United States put bombers loaded with H-bombs in the air twenty-four hours a day.*

The shock of Sputnik began setting in as it became apparent that the United States was not going to answer Sputnik with a satellite of its own in the next few weeks. The nation had been beaten, and all that the people at the highest level of leadership could do was to tell Americans that it was not all that important. Privately, however, White House advisers admitted that the first full week following Sputnik's launch was one "prolonged nightmare," with groups from Congress, the Pentagon, and Capitol Hill dashing in and out of the president's office.

The calm at the center of this whirlwind was Dwight David Eisenhower. Throughout the month of October he maintained his composure, setting the tone on the ninth with a White House press release in which he congratulated the Soviet scientists on Sputnik.

The previous year, Eisenhower had been reelected president of the United States by a landslide, carrying the electoral college with the votes of

*Bombers would stay in the air twenty-four hours a day until they were grounded in the 1970s when rapid launch ground alert was substituted.

forty-one of the forty-eight states. He had won his first term in the White House on the simple and effective slogan "I like Ike" and was, in his fifth year in office, as close to having universal admiration as any twentieth-century president. The World War II leader, who had brought the troops home from Korea and heroically fought back from a first-term heart attack, won on the promise that he could bring the nation four more years of peace, prosperity, and progress.

The message about Sputnik presented to the public by the president's men was simply "don't worry, don't panic." Outgoing secretary of defense Charles E. Wilson dismissed the launch by saying, "Nobody is going to drop anything down on you from a satellite while you are asleep, so don't worry about it." The secretary of the treasury, George Humphrey, cautioned, "The real danger of the Sputnik is that some too-eager people may demand hasty action regardless of cost in an attempt to surpass what the Russians have done." Rear Admiral Rawson Bennett, of the Office of Naval Research, tossed it off as "a hunk of iron almost anybody could launch."

Percival Brundage, director of the White House's Bureau of the Budget, glibly remarked at a Washington dinner party that Sputnik would be forgotten in six months. Perle Mesta, the famous Washington hostess, heard the remark and added, "Yes, dear, and in six months we may be dead."

But over the next weeks and months not only was Sputnik not "forgotten," it seemed to become more important. And the more Americans were told by the men in Washington not to worry, the more they panicked. For all its simplicity, small size, and inability to do more than orbit the Earth and transmit meaningless radio blips, Sputnik's impact on America and the world was enormous and totally unanticipated. To the man or woman in the street, it was vastly confusing and most threatening. Some polls did not register panic, such as one by Samuel Lubell conducted in New Jersey, which showed an outward lack of concern but reported misgivings and the need for America to catch up.

"We didn't know what they were talking about. Something up in orbit? Launch stuff up into the upper atmosphere? We didn't know what Sputnik was," said Joe McRoberts, then working as a civil-defense administrator in Michigan and later news chief for the Goddard Space Flight Center. "All the attention at the time had been on missiles, and we seemed to be doing very well there. And then, all of a sudden, here's this thing going round and round the Earth and not coming back down. All of your life you've been taught that what goes up must come down."

The United States was ill prepared for the shock of Sputnik both "patriotically and psychologically," in the words of William Shelton, for many years the expert on Soviet space programs at the Library of Congress. The result was fear.

There was actually more than one fear at work. The first and most universal may have been fear of the unknown—fed by wild, often unattributed reports granting extraordinary powers to the satellite. For instance, it was "presumed" in one Associated Press story that the U.S. armed services were hiding aircraft, tanks, and other hardware so that Sputnik spy equipment would not be able to sense them and send back an inventory of American military holdings. Of the many bits of speculation as to the next Russian move, none was quite so dramatic as that of John Rinehart of the Smithsonian Astrophysical Observatory in Cambridge, Massachusetts: "I would not be surprised," he said, "if the Russians reached the moon within a week."

Feeding the fear was the fact that nobody really knew what Sputnik was doing up there or at first even what it looked like. Many American newspapers, lacking an initial image in the first days, published artists' renderings of the Navy's Vanguard satellite that had been widely distributed earlier in the year. Using the American satellite's image to depict the Russian satellite added to the humiliation and overall confusion.* Susan Cottler, a professor of history at Westminster College in Salt Lake City, recalled this confusion in a National Public Radio interview on Sputnik's fortieth anniversary: "Everyday people—teenagers, people in their 20s, parents—would look up in the sky at night and try to spot it. They would wonder if they had to speak in hushed tones, or have arguments behind closed blinds, for fear that Big Brother was, in fact, spying on us."†

Even Sputnik's simple beeps brought confusion, which fed the fear. The CIA, Defense Intelligence Agency, Army, Air Force, and other Western intelligence worked around the clock to see who would be first to decipher the beeps. The assumption had been that important data were coming back from space. Some went out on a limb. Columnist Stewart Alsop suggested on October 13 that the Soviets had actually put a reconnaissance satellite in orbit. "There is a mounting body of evidence, taken most seriously in the

Life magazine staffers wanted to know where to send an airplane in order to photograph Sputnik. "Cost is no object!" was what they told J. Allen Hynek of the Smithsonian Astrophysical Observatory, who said later, "If I had said Tierra del Fuego, within minutes, the *Life* crew would have been dispatched there."

†She added that Sputnik inspired paranoia on the silver screen in such films as *On the Beach* (1959) and *The Manchurian Candidate* (1962).

Washington intelligence community, that the Soviet satellite is not blind," he wrote, "that Sputnik has eyes to see." Finally, one of the Russian IGY delegates in Washington revealed that there was no code and the satellite could not see. The one-watt, battery-operated transmitter was placed inside the aluminum shell simply so that it could be tracked. Few paid attention to these comforting words, and those who did would not believe them. "Many believe that the whole story has not been told" was *Time* magazine's response. "The CIA had better get to the bottom of it, the man on the street muttered," William Manchester wrote in *The Glory and the Dream*, "or the U.S. taxpayer would know the reason why."

So convinced were some American citizens that the Russians were up to something that they signed on as unappointed advisers eager to help interpret Sputnik's secret mission. Wallace A. Bounds of the Rockwell Manufacturing Company of Sulphur Springs, Texas, wrote to the secretary of defense on October 19 pointing out that Sputnik was probably collecting coded data, while expressing confidence that the government already knew this. He suggested that "to break Sputnik's code—if it happens to be mapping the earth by some means, possibly infrared—we can compare its oscilloscope patterns for selected given positions with actual geographic features. . . . If this is so and they are mapping us, we can also map them if we have been recording its signals when it is over Russia." Bounds went on to argue that even if the United States could not break the code, it should record and preserve everything for later interpretation. The letter and others like it were taken seriously and kept from public view.

Adding to this fear was what old-time newspaper folks refer to as a "silly season"—when the most outlandish happenstances and claims become newsworthy. Only this time it was different, for the silly season had an odd, threatening edge to it. Attention was paid to minor hoaxes, such as bogus radio signals sent out by miscreants with transmitters, the 1950s' equivalent of today's computer hackers.

All sorts of mysterious sightings became newsworthy that at another time would have been shrugged off. Even the *New York Times*, in a particularly dull period, dutifully reported such diverse items as a takeover of police radio transmissions in Stockholm and a claim by "some scientists" in Australia that the Russians had a radio transmitter in Antarctica "questioning" the satellite each time it passed over their country. As far as the *Baltimore Sun* was concerned, everything was worth reporting, even banquet prayers. Maryland governor Theodore R. McKeldin, addressing the Liberty Republican Club at the Lord Baltimore Hotel in mid-October, said: "We are gathered here under the sign of Sputnik. Let us not be distracted from the joys, the hopes, the duties and responsibility of free men and women by the 'beep-beep' of a little moon that is soon to vanish in a fast descent."

All the stranger was the fact that some of the stories actually were true. When the Associated Press reported on a Schenectady, New York, doctor who claimed that Sputnik was opening his garage door, similar reports flooded newspapers all across the country. Some were obviously false—such as those reporting that the doors opened the same time each day—but others were real. John M. Williams, a hurricane researcher at the Florida Institute of Technology in Melbourne, Florida, was working for Collins Radio at Cape Canaveral in 1957 and had his electronic garage doors opened three times by Sputnik. "The first time it happened about 2 A.M. and it scared me to death. I grabbed my .45 because I figured that someone was coming after my car."

In some cases, religious voices added to the confusion and the fear. At first, the semi-official Vatican newspaper, *L'Osservatore*, stated that "God has no intention of setting a limit to the efforts of man to conquer space." Five days after the launch, the official Vatican position was that Sputnik was a "frightening toy in the hands of childlike men who are without religion or morals."

On the second Sunday after Sputnik—October 13—the pulpits of America rang with every sort of commentary, a few going so far as to assert that it foretold the Second Coming of Christ. "Don't be surprised, my friends, if He comes today," a minister was heard to say to a congregation that included a reporter from the *Washington Post*. Atheists, on the other hand, somehow reasoned that it proved that God did not exist. Joseph Lewis, president of the Freethinkers of America, crowed, "The new earth satellite . . . broadcasts no discovery of God in the heavens. . . . What a mockery does this great scientific achievement of man make of the petty religions of the Earth." An East German newspaper noted that the Russian scientists needed no God in their equations to get into space and that this was somehow proof of atheism.*

With a handful of notable exceptions, the press was totally unprepared

*Perhaps the oddest spiritual impact of Sputnik took place when Little Richard saw Sputnik in the sky while performing on stage in Sydney during an Australian tour. Taking it as a divine sign, the singer marched off the stage, renounced rock 'n' roll, and—for a while at least—became an evangelist.

for Sputnik and the perpetual misinformation relating to it. Reporters whose beat on Friday morning had been crime or city hall were handed the Sputnik assignment on Friday night. "Misquotes were the rule; sensationalism rampant," observed Eloise Engle and K. H. Drummond in their book *Sky Rangers*. Many reporters lacked sources for researching the story. Even with good sources, newspaper stories sometimes got twisted. At the major clearinghouse for information, the Smithsonian Astrophysical Observatory in Cambridge, Massachusetts, the simplest statement was occasionally transformed into a lurid headline. For example, a statement was released by its director, J. Allen Hynek, to the effect that the Sputnik orbit could not be explained solely by Newton's laws but that other factors such as atmospheric drag were involved. The story was published in a Boston paper with the melodramatic and totally misleading headline "Mysterious Force Grips Sputnik."

There's no question that the most basic fear instilled in Americans by Sputnik was fear for their lives. They were coached to be afraid by the press and the politicians. Senator Mike Mansfield, a man not known for overstatement, examined the situation and said, "What is at stake is nothing less than our survival."

Sputnik became much more than a simple scientific exercise. It was a vivid missile display. Suddenly, earlier Soviet claims of missile power—claims discounted publicly by officials in the West—were being taken in deadly earnest. Six weeks before the launch, the Kremlin had announced its new super-long-distance intercontinental multistage missile, saying the Soviet Union was now able "to direct rockets into any part of the world." Now the warning took on a chilling immediacy, especially when the word *missile* was used in place of the word *rocket*.

If the Soviets could send a satellite spinning around the globe every ninety-six minutes, it took no great feat of imagination to believe that they could loft some more menacing projectile toward the United States. Americans became convinced that the science and math that had put this object into a seemingly perfect elliptical orbit could just as easily target any site on Earth with a nuclear warhead. No less a figure than Senator Richard Russell, chairman of the Senate Armed Services Committee, was among the first to give voice to that belief. The newspapers caught on quickly. The fear caused by that rocket power—the ability to "brute-force" a crude, heavy satellite into orbit—was expressed in an October 7, 1957, editorial in the *Chicago Daily News*: "If [the Russians] can deliver a 184-pound 'moon' into a predetermined pattern 560 miles out in space, the day is not far distant when they could deliver a death-dealing warhead onto a predetermined target almost anywhere on the earth's surface."

This particular fear was well nourished as other politicians, scientists, and public figures seemed to be competing to see who could create the greatest general hysteria. It began to dawn on America that perhaps this was not just a scientific event but a harbinger of intercontinental ballistic warfare and, even more threatening, war waged from space. Some asked if the next Sputnik would have atomic bombs aboard. On the second Monday after Sputnik, *Newsweek* suggested that a couple dozen of the Sputnik launch vehicles "equipped with dirty H-bombs instead of radio transmitters and batteries, could with very few technical changes be made to spew their lethal fallout over the U.S. or Europe." Senator Lyndon Johnson said of the Russians, "Soon, they will be dropping bombs on us from space like kids dropping rocks onto cars from freeway overpasses."

The result was that Americans were scared half to death and people had panic reactions. Even admonitions not to worry tended to have an ominous ring. From his home in Independence, Missouri, Harry Truman told the American people "not to panic"—and then went on to discuss the satellite's dire military implications.

Edward Teller, then associate director of the Livermore Radiation Laboratory and a father of the hydrogen bomb, went on television to suggest that the future might now belong to the Russians and to proclaim that the United States had "lost a battle more important and greater than Pearl Harbor." Labor leader Walter Reuther called it a "bloodless Pearl Harbor." The comparison was everywhere, particularly in Washington. "[There] was a sense of foreboding that the city had not known since Pearl Harbor," *Washington Post* writer Edward T. Folliard recalled a year later. There were other historic comparisons that emphasized the enormity of the event. To the *New Republic* the launch of Sputnik was likened to the discovery of America by Christopher Columbus; *Life* compared it to the first shot at Lexington and Concord, urging Americans to respond like the Minutemen.

Sputnik clearly undermined America's defense, which in 1957 was based on a simple policy of deterrence, officially defined as "being so obviously superior in our ability to carry the war to an enemy that he will not take the risk of starting one." A central tenet of deterrence was that the United States had to remain technologically superior to any potential enemy. "America is worried," wrote financier and presidential adviser

Bernard Baruch. "It should be; we have been set back severely, not only in matters of defense and security, but in the contest for the support and confidence of the peoples throughout the world."

The President Under Pressure

It had long been taken for granted in American scientific and military circles that when time and technology were ripe, the United States would head into space—first. Before the Sputnik launch, this view was widely shared by the American public. The Space Age, they were told time and again in their magazines and newspapers, would dawn on the flats of the Air Force Eastern Test Range at Cape Canaveral, Florida, with American radios and telescopes properly trained to the great event. Editorialists had asked: Hadn't the Germans' inability to produce an atomic bomb during World War II proved the innate superiority of "free" science over totalitarian science?

Now that the Russians had launched first, America had to adjust to its new status and uncertain future. The Democrats were generally more critical than the Republicans. Democratic senator Henry M. Jackson of Washington termed Sputnik "a devastating blow to the prestige of the United States as the leader in the scientific and technical world" and demanded a "National Week of Shame and Danger." Senator John F. Kennedy, who would later make much of the loss of American prestige through Sputnik in his 1960 presidential campaign, told an audience in Albuquerque that because of Sputnik "the impression began to move around the world that the Soviet Union was on the march, that it had definite goals, that it knew how to accomplish them, that it was moving and that we were standing still. That is what we have to overcome."

President Eisenhower's political enemies made hay out of his seeming lack of concern for Sputnik. The Democrats, seizing a priceless opportunity, denounced what Hubert Humphrey called this "pseudo-optimism" from 1600 Pennsylvania Avenue.

Indeed, Eisenhower's administration was making a concerted effort to keep the word *space* out of the vocabulary of American government and industry. It was as if not uttering the word was enough to keep the problem under control. For example, Simon Ramo, of the Ramo-Woolridge Company, later to become TRW, had registered the name Space Technology Laboratories with the trademark office in the mid-1950s, anticipating its use someday. On the Monday after the Sputnik launching, he used it to rename the company's

Guided Missile Research Division, but got a call from the deputy secretary of defense within days asking him to consider changing it back. Ramo was told that "this little basketball of the Russians" would be forgotten in two weeks. Ramo, who felt the launch was an event in the same class as Lindbergh's flight, "considered" the change for the next two weeks before rejecting it and never heard another word from the Pentagon on the matter.*

The administration's reluctance to talk about space was not shared by others—especially the press. When Eisenhower went before the press on October 9, 1957, five days after the launch, virtually every question dealt with the ramifications of the satellite spinning several hundred miles overhead. It was "the most skeptical, if not overtly hostile press conference of his presidency," according to biographer Geoffrey Perret.

Eisenhower said the Soviets had "put one small ball in the air," adding, "I wouldn't believe that at this moment you have to fear the intelligence aspects of this." The comment that got the most attention was his insistence that the Soviet satellite had not raised his apprehension "one iota." He acknowledged that Sputnik's weight "has astonished our scientists," and granted that the Soviets might have gained "a great psychological advantage" throughout the world, but asserted that German science and scientists, not Russian, were behind it.

In the weeks immediately following the launch, many members of Eisenhower's inner circle made dismissive Sputnik comments, including presidential assistant and chief of staff Sherman Adams. He wrote it off as "one shot in an outer-space basketball game." The basketball metaphor was—and still is—the line most often recalled when attempting to show the degree to which Eisenhower's inner circle seemed to be out of touch with the realities of the situation. Richard M. Nixon later wrote in his memoirs, "I believed that this flippant remark was wrong in substance and disastrous in terms of public opinion."†

*The Space Technology Laboratories became the first facility in American industry to build spacecraft. Pioneer I was its first spacecraft, and Pioneer 10 the first body to leave the solar system. Ramo, eighty-six, told this story when being honored by NASA at a symposium on November 5, 1999. Pioneer 10 continues to send back signals to Earth. On April 21, 2001, as this book was going to press, a station in Madrid picked up a Pioneer 10 signal from a distance of 7.30 billion miles.

†In 1974 White House aide Bryce N. Harlow admitted in an interview with a NASA historian that he had written the basketball game line for Adams and insisted that it really was reflective of the thinking inside the White House at that time. He said of his line, "It wasn't clever, but it was cute."

President Eisenhower under fire from the press at his October 9, 1957, press conference.
(*U.S. News & World Report* Collection, the Library of Congress)

In general, media commentators, the public, and many on Capitol Hill found the president's response to Sputnik and his reassurances on the military significance of Sputnik inadequate. Doris Fleeson, a *Washington Evening Star* columnist, wrote that Eisenhower gave the satellite the "Miltown treatment," alluding to the Prozac of the 1950s. Democratic senator Stuart Symington declared that the president was "paternalistically vague." Others were harsher. Walter Lippmann, perhaps the most influential columnist of the time, held that the president was "in a kind of partial retirement" and letting the country drift and decline. Edward R. Murrow of CBS was talking about Eisenhower and his administration when he said "the key men in Washington had not the imagination to understand what it would mean for the Soviet Union to launch its satellite first."

President Eisenhower was taking it on the chin, and his defenders were few. One of them was David Lawrence, editor of *U.S. News & World Report*, who said that the president's refusal to be stampeded was an example of "courageous statesmanship."

On October 10, six days after the launch, there was a special meeting of the National Security Council, held to address the issue of the implications of Sputnik on U.S. security. CIA director Allen Dulles began by acknowledging that the actual launching of the Earth satellite had not come as a surprise to the intelligence community. In fact, he predicted that there would be an additional six to thirteen Sputniks before the IGY came to a close at the end of 1958.

Dulles saw Sputnik as part of a "trilogy of propaganda moves," the other two being the announcement of the successful testing of an ICBM in August and the recent test of a large-scale hydrogen bomb, staged by Khrushchev to impress the underdeveloped world. He went on to say that the Russians were getting maximum propaganda value out of these events.

At the conclusion of Dulles's generally despairing briefing, Assistant Secretary of Defense Quarles was asked to speak. After outlining the development of America's satellite programs after World War II, Quarles articulated the administration's policy on Vanguard not interfering with high-priority development of ballistic missiles. He emphasized that the Russians had done the United States a "good turn" by launching Sputnik, thus "establishing the concept of freedom of space." America's first big prize from space was going to be its ability through future reconnaissance satellites to count Soviet missiles and aircraft on the ground.

(In fact, one day earlier Eisenhower had received a memo from the outgoing secretary of defense on the very same subject. Simply labeled "Earth Satellite," the memo made the point as clearly as possible why it was imperative that the United States stick to the plan of a scientific, nonmilitary satellite: "We in Defense were concerned at that time about international reactions to a reconnaissance satellite that the Air Force was giving serious study to. It was felt that scientific satellites which would be clearly nonmilitary and clearly inoffensive might help to establish the principle that outer space was international space. Thus reconnaissance satellites traveling in it could not be objected to by the countries overflown because the space is free and the satellite itself is inoffensive in character.")*

On the topic of spy satellites, the president expressed his dissatisfaction over a newspaper report that two "so-called intelligence people" had claimed Sputnik was actually taking reconnaissance photographs of the United States for the use of the Soviet Union.

Who precisely, Eisenhower wanted to know, was doing this kind of talking?

Assistant Secretary Quarles replied that he did not know the two individuals in question and would probably never find out who they were. Neverthe-

*This memo, which did not lose its Top Secret classification until 1998, indicated quite clearly that when the IGY satellite resolution was adopted by both the United States and the USSR in October 1954, the Department of Defense assumed that the Russian scientists were innocently concurring.

less, he doubted the truth of any such rumors, though the U.S. government could not know for certain that the Soviet Earth satellite could not take pictures. In any case, he couldn't conceive of any knowledgeable person in government making such an allegation as this, which, to the best of his belief, was groundless. For the National Security Council, the less talk about satellites taking pictures from space, the better.

Others were as reassuring as Quarles. Alan Waterman, head of the National Science Foundation, told the president, his cabinet, and top military leaders that Sputnik introduced no military surprise or unexpected threat.

Eisenhower then calmly explained his position. He admitted that the Russians had gained a clear propaganda advantage, but that the United States had a sound plan and he wanted to stick with it. The plan was to continue with scientific satellites, while putting the real effort into missiles, military hardware, and spy satellites.

Later that day, in a meeting with his staff, Eisenhower compared his plan for space to the Anglo-American plan for defeating Nazi Germany in World War II. The military plan was under constant attack. "We never did abandon it," said Eisenhower. "It was a good plan, a long-range plan that had been carefully worked out. We went on and won."

The difference between World War II and the Sputnik crisis was that this time some of his harshest critics were men in uniform—his own generals—who in the weeks ahead would take him and the plan to task in the press and in witness chairs on Capitol Hill.

Generals in Revolt

At the time of the Sputnik launch, the Army's power and presence had been slipping since its days of glory in World War II. Wernher von Braun and company were now the Army's best shot at regaining some of the lost glory through missiles and space, which explained the intensity of the Army's determination to make an issue out of Sputnik and the selection of Vanguard.

Generals in the Pentagon and at Huntsville, and their German scientists, started to vent their long-standing frustration over the fact that the Army had been stopped from sending up a satellite in 1956. Two notable examples were Major General Holgar Nelson Toftoy, the man who had been instrumental in getting the Rocket Team out of Germany and who was then head of the Redstone Arsenal in Huntsville, and Brigadier General John A. Barclay, his deputy commander. They used as their platform the International Astronautical Congress in Barcelona, Spain, in October. Petulantly, they charged that they had been meddled with by their superiors, whom they did not name but who were clearly the president, the secretary of defense, and the secretary of the Army. They held nothing back in their open disdain for this situation.

This undisciplined outburst at an international event did not sit well with Washington, especially with the president, who received a clipping of the AP story the next morning. It was the same morning that he swore in Neil McElroy as his new secretary of defense. Once the ceremony and the coffee-and-cake reception were over, the president called all the top Pentagon officials into the Oval Office for a closed-door tongue-lashing. Eisenhower was furious that his own generals could not understand that the decision to separate the military missile and rocket programs from the civilian IGY program was made on purpose and for good reason. "When military people begin to talk about this matter and to assert that other missiles could have been used to launch a U.S. satellite sooner," said the president, "they tend to make the matter look like a race, which is exactly the wrong impression. I want to enlist the efforts of the whole group on behalf of, 'No Comment,' on this development."

Stern, unequivocal orders were immediately issued by the secretary of the Army for the Army to be quiet. It was Ike's turn to throw a punch. This was the same day, October 9, of the aforementioned press conference at which Eisenhower asserted that the reason the Russians had gone first was that they had "captured all the German scientists" in Peenemünde. The Huntsville team was dumbfounded and the local press was shocked by the president's remark. The *Huntsville Times* Washington correspondent wrote, "By his words at his weekly press conference, the President indicated he was unaware of the world-famous team of former German scientists now at the Army's Redstone Arsenal at Huntsville, Alabama." Neither Eisenhower nor press secretary James Hagerty was willing to comment on the matter, according to the *Times*, which pointed out with alarm that Hagerty was in New York City attending the World Series. Of course, Eisenhower certainly knew about the Huntsville Rocket Team; therefore, the comment and the unwillingness to correct it would appear to be part of a not-so-subtle payback for the direct assault by von Braun.

In truth, the United States had gotten the most talented Germans. The Russians had launched Sputnik on their own, with the help of some German technicians and V-2 production rocket staff. Both nations had borrowed heavily from the V-2. But unlike the United States, the Soviet Union did not have any leading V-2 experts; nor had it captured any operational V-2s after the war.

Huntsville was so incensed by Eisenhower's comment that the next day the spirit of the gag order was disobeyed. A reporter named Steve Yates of the North American Newspaper Alliance filed a story from Huntsville quoting a "top missile man whose name is nationally known" as saying that their pleas to help the Vanguard program have been thwarted and that they could get a satellite in orbit in three weeks to six months. His source was most certainly Wernher von Braun. Although a civilian, von Braun was in the employ of the Army and thus subject to the gag order. He got around this by directing Yates to earlier quotes. For instance, the article cited an old 1954 quote from the *Birmingham News* in which von Braun had said, "Give me five years and $5 billion and we can land on the moon."*

Throughout the month the story of the thwarted team in Huntsville was repeated in numerous newspaper articles that cast blame on a host of people, up to and including the president. On October 25 investigative columnist Drew Pearson revealed that "six satellites," all ready to launch, were gathering dust in Huntsville. (They were not satellites but actually Jupiter C rockets.)

As a result of public interest in the Huntsville team, von Braun now became America's space star, even comfortable telling anecdotes about life in the Third Reich. In a *New York Times Magazine* article that appeared on October 20, he reminisced about his disagreement with Heinrich Himmler over the suggestion to speed up the production of V-2s. "I remembered he was interested in horticulture, and I said that it was all right to put a plant into the ground and carefully nurture it, but that too much manure, for example, could kill it. Himmler smiled weakly, but I thought little of it until I was arrested three weeks later by the Gestapo. It took a direct order from Hitler to release me."† He stressed his quick as-

*In fact, von Braun and company continued complaining bitterly to anyone who would listen that a terrible error had been made. William Tucker of the UP was told by the crew in Huntsville that their request to launch had been turned down no fewer than five times. Tucker reported that the Huntsville team "labors on and nurses its bitterness."

†That von Braun talked about Hitler in this interview was highly unusual. Yet his comments were hardly damning as he did little more than brand Hitler an atheist: "I met Hitler four times. My first impression was that here was another Napoleon, another colossal figure who had upset the world. But at my last meetings with him—when I explained some of the technicalities of the V-2—Hitler suddenly struck me as an unreligious man, a man who did not feel that he was answerable to anyone, that there was no God for him."

Wernher von Braun answering questions at a press conference on February 1, 1958. (*U.S. News & World Report* Collection, the Library of Congress)

similation in the United States, referring, for example, to his American-born daughters as his "little Texan" and his "Alabama belle." The *Times Magazine* called him the "Prophet of the Space Age."

Von Braun was featured in *Life* magazine a few weeks later and was portrayed as a martyred hero whose early bids to put up satellites were "usually dismissed as if made by a tiresome crackpot" and whose prophecies were "unheeded." *Life*'s headline referred to him as "The Seer of Space."

With the assent of the Army, von Braun was becoming more outspoken by the day. He was speaking easily of doom one minute ("A man in Sputnik means control of the globe," he told the *New York Mirror*) and scolding the president the next.

On October 29, 1957, von Braun gave a secret briefing to Army officials in Washington, titled "The Lessons of 'Sputnik,' " in which he saw the event as a "national tragedy" that had done great harm to American prestige around the world. That said, he went on to make some stunning points that the brass almost certainly were not prepared to hear. The first was that the United States was committing a grave error in not being able to appraise the research and development capabilities of a nation run by a totalitarian regime. As one having had "the dubious privilege of living and working under a totalitarian government for many years," he disputed the notion that captive science was science in terror. Next on his agenda was to point out how scattered and counterproductive American military research and development had become: "About a year ago I saw a compilation of all guided missile projects, which—at one time or another—had been activated in this country since 1945. I doubt if you will believe it, but the total figure was 119 different guided missile projects!"

He then explained that these were not paper studies but full-fledged development projects, only a few of which had any promise of success. America, he said, needed to forge teams to tackle the problems of missiles and space. He told the assembled group, which included the secretary of the Army, that Americans understood the concept of teamwork when it came to football and baseball, but not when it came to the very technology that could save the West.

He proceeded to tell those assembled that the Russians were about to rack up more space successes and "firsts" and that the public had to be prepared. Even if the United States started to catch up immediately, he said, the previous delays would still cost the nation years of "helplessly witness[ing] how the Soviets, in full view of millions watching in awe, reap the easy fruits of the first conqueror of outer space."

Domestic Troubles

Guessing what the Russians would do next became something of a national obsession. Speculation centered on the Moon. One widely published prediction held that the anniversary-minded Russians might want to mark the fortieth anniversary of their Revolution on November 7 by aiming an unmanned spacecraft for the Moon and marking its subsequent arrival with the detonation of an A-bomb. An article in the *New York Times* speculated that an unmanned November Moon shot would land and discharge "a red paint or dust that would mark an area large enough to be seen from the earth during the full moon."

The idea had taken hold that if the Russians got to the Moon first, they would own it. At the aforementioned October meeting of the International Astronautical Congress in Barcelona, an American lawyer by the name of Andrew G. Haley said that present international law would give Russia the right to claim the Moon if it became the first nation to hit it with a missile or land men there.

A 1957 *Washington Post* editorial likened the state of America at the time to the nation on the eve of World War II: "To overcome the scientific and psychological disadvantage, it will be necessary for the Administration first to assert some aggressive leadership and then to provide whatever talent and money are necessary to right the balance."

The balance was important, but Eisenhower had other things on his mind besides Sputnik. At the top of his domestic list was the crisis in Little

Rock, Arkansas. On September 2, 1957, segregationist governor Orval Faubus had called out the Arkansas National Guard, ordering them to surround a white school in Little Rock to "prevent any violence"—a euphemism for blocking eight black students from entering the building. On September 24 the students finally entered the school, but threat of a riot forced the school to evacuate the black students. The next day, the children were escorted into the school by President Eisenhower's federal troops, who remained for the rest of the school year.

Eisenhower also had to contend with a faltering economy in the United States. On Monday morning, October 21, the bottom had dropped out of the stock market when it experienced its largest one-day loss in two years. Fear of Sputnik and uncertainty were undercutting the market, and it dropped again the next day. Sputnik did not start the slide but exacerbated it. The Dow had dropped 21 percent in value since July 12, and the bull market that had been in place for more than three years was now over. A recession was in full swing; personal and business income were both down for the year, and unemployment was on the rise. Speculation on Wall Street was that the president deliberately had not reacted strongly to Sputnik to minimize its economic impact. It had been argued that if Eisenhower had expressed fear and panic, there would have been a run on the banks.

Although these were volatile domestic concerns, Eisenhower's primary worry was the defense of America. And he was commander in chief of armed services that were in a bizarre, self-defeating dogfight for control of missiles and space—a dogfight which had been going on for years. The fiercest fight was between the Army and the Air Force. The conflict could be boiled down to their different concepts of the brute force needed to launch satellites. The Army believed that intermediate- and long-range missiles were guns or cannons, while the Air Force saw them as aircraft. For this reason, the Army called them rockets, and the Air Force called them missiles. Their unresolved friction had cost the nation millions and would continue to do so; it was also responsible for the public's enduring distrust of the military.

Seeing and Hearing Sputnik

As October wore on, Americans' apprehension was reinforced by the fact that the Russian spacecraft was actually passing over the United States four to six times a day! For twenty-one days Sputnik continued to send radio signals, which were being monitored by ham radio operators and occasionally broadcast on network radio. To many, the fear was purely territorial: There it was overhead—visible to the naked eye and audible to anyone with a shortwave receiver. America had fought two world wars, protected by the breadth of oceans and the comfort of a strong Navy. A certain sense of invulnerability seemed to be an American birthright. Ever since British troops had sacked and burned most of Washington, D.C.'s public buildings, including the Capitol and the president's house on August 24, 1814, the United States had done everything it could to isolate itself from foreign invasion; all subsequent generations of Americans had felt they were safe at home.*

The skies over the continental United States had *never* been violated—during two world wars not a single enemy aircraft had penetrated mainland airspace. Now suddenly an object controlled by a hostile power was directly overhead. The fact that Sputnik could be seen and heard was all-important to its dramatic effect. It was visible only in the morning or in evening twilight, and in the United States it sometimes showed up just as the evening news came on television. Network executives were stunned by the fact that on some evenings much of their audience simply walked out into the backyard to see history with their own eyes.

Five days after the launch, the Soviet Union released a photograph of the actual satellite posed on a three-legged stand. It was a simple, shiny ball, with four trailing antennae, which looked to many like a piece of modern sculpture.

Astrophysicist Donald D. Clayton was studying for his doctorate at the California Institute of Technology when a physics colloquium was interrupted for a sunset look at Sputnik. As he recalls in *The Dark Night Sky*: "As a body, 200 physicists—eminent professors, postdoctoral assistants, and students—filed to the roof of Bridge Laboratory to watch for it. Many sudden shouts and pointing fingers later, it came like speeding Venus across the twilight. The chatter died quickly away, and in five minutes of silence we watched it pass. Just as suddenly, it was gone, and we all stared dumbly at each other, just barely comprehending that mankind would never be the same again."

*For example, on January 27, 1838, a quarter century after the British invasion, Abraham Lincoln asked: "At what point shall we expect the danger? By what means shall we fortify against it? Shall we expect some transatlantic military giant to step the Ocean and crush us at a blow? Never! All the armies of Europe, Asia and Africa combined with all the troubles of the earth . . . could not by force take a drink from the Ohio, or make a track on the Blue Ridge, in a trial of a thousand years."

Because of its orbital pattern, Sputnik debuted on different dates in different places in North America. At dawn on October 13, Sputnik and its rocket were first visible over Manhattan. As it crossed New York's Central Park for the first time, NBC filmed it, and later that day Sputnik made its first appearance on network television. Even the most skeptical New Yorker could no longer shrug it off.

The sound coming down from orbit, the first man-made sound from the heavens, was Sputnik transmitting for twenty-three days before the silver-zinc batteries ran down. Sputnik had dual transmitters on frequencies of 20.07 and 40.002 megahertz, with sound radiated by four external, spring-loaded whip antennae. The beeps were alternately transmitted on each of the two frequencies. Each beep was 0.3 seconds in duration with a 0.3-second pause. During the pause on one frequency, the signal was transmitted on the other frequency.

The frequency of the "beep-beep" transmission was between two easily accessed ham radio bands. Amateur radio operator Roy Welch was quick to record the beeping signal from his home station in Dallas. A few days later, on October 9, he was playing the recorded beeping for long lines of interested visitors to the Texas State Fair. In New York City the sound of Sputnik was the hit of the high-fidelity show, where it was also rendered graphically on oscilloscopes.

It was a popular belief at the time that the Soviets chose the lower frequencies to allow ham radio operators around the world to receive the signals and track the satellite. This scheme, it was thought, not only popularized the project but also provided cheap worldwide tracking reports from hams around the globe. In 1990 Konstantin Gringauz, the man who designed the transmitter, confirmed the suspicion in an interview when he told Brian Harvey of *Spaceflight* magazine, "Korolev was adamant that signals should be received by as many people as possible throughout the world, including amateur radio hams." In addition, the Soviets had gone out of their way to make the frequencies known. As Gringauz reminded Harvey: "In the summer of 1957, announcements were made of the impending launch of Sputnik 1. The June 1957 issue of *Radio* magazine, which had a circulation of over one million, published a complete description of . . . its transmitters, and how to pick it up, but no one took it seriously."*

Remarkably, among those who did not take these Russian announce-

*One exception was the Japanese Radio Operators' League, whose members tracked Sputnik from the moment it went up.

ments seriously was the Vanguard team. A little-remembered and amazing fact is that the worldwide Navy tracking system, which had been put into operation three days earlier, did not work on Sputnik's frequency. The American "Minitrack" antenna system for tracking the International Geophysical Year (IGY) satellites had gone into operation three days before Sputnik was launched. Minitrack was designed for the agreed-upon frequency of 108 megahertz, so it was on the wrong station. American scientists complained to the newspapers. A small story in the *Washington Evening Star* the day after the launch was headlined "Russians Tricked U.S. Scientists on U.S. Megacycle Band." Forgotten in all of this—perhaps on purpose—was that American scientists had been given the proper frequencies by V. A. Kotelnikov on August 27, 1957, at a conference in Boulder, Colorado.

As a result of Sputnik, amateur radio operators in the United States began talking to their counterparts in Russia. It was all quite civil and friendly, as these "hams" had a powerful avocation as a common interest. The Russians were proud, while the Americans tended to be congratulatory. "I told Yura what a wonderful job the Russians are doing in research," a Virginia operator calling himself "Sparky" told the *Washington Post*, referring to a radio buddy in Odessa. In Washington, the State Department made no secret of the fact that it was uncomfortable with this chatty, grassroots diplomacy.

World Reaction

If Eisenhower had been unprepared for the reaction to Sputnik, so was his counterpart in the Soviet Union. On first hearing the news of Sputnik's launch, Nikita Khrushchev had been noncommittal about "just another Korolev rocket launch," according to Korolev's former colleague Boris Rauschenbakh. But as Khrushchev became aware of the degree to which the West was responding to Sputnik, he granted an interview to James Reston of the *New York Times*. "It must be realized that the Soviet Union is no longer a peasant country," he proclaimed triumphantly. He told Reston that Soviet missiles were now plentiful and came in every size and shape, while the piloted long-range bomber favored by America was becoming a thing of the past.

The Soviets were far more sophisticated in the art of public relations than the West had realized. Soviet scientists were placed in key places around the globe to maximize the impact of Sputnik. Predictably, the Russian press tried to seize the advantage in the battle for international sway. A *Pravda* dispatch from New York claimed that certain U.S. senators were "showing signs of hysteria." But even among America's staunchest allies, there was doubt about its ability to bounce back from the blow dealt by Sputnik. The *Times* of London wrote of "the demon of inferiority which, since October 4, 1957 . . . has disturbed American well being."

If a demon of inferiority had indeed been unleashed, it was now being well fed by a steady stream of reports from abroad. The most common reaction in Europe seemed to be that Sputnik was a warning to the United States and other Western nations about Soviet missile power. To editorialists in Madrid and Rome, it meant that Russia had the means to send nuclear missiles to any point in the world. Confidential government opinion surveys, which were to remain classified for a number of years, showed that over 90 percent of those surveyed in western Europe had heard of the achievement and the same percentage knew it was a Russian feat—unprecedented levels of awareness in this kind of research—while enthusiasm for NATO and the Western alliance was diminishing with striking speed in France and Italy.

A Gallup poll discovered that U.S. prestige had eroded in six of the seven foreign cities included in its survey, and within weeks there was a decline in public enthusiasm for "siding with the U.S." and NATO in Germany, France, and Italy. Unlikely foreign leaders began praising the Soviet Union and questioning the scientific ability of the United States. Of these, the most surprising was Spain's Generalissimo Francisco Franco, who was making unprecedented pro-Soviet and anti-American remarks in the wake of Sputnik. By any measure, America's image had been damaged abroad, and the word was getting back to the United States.

Less than a week after the launch, on October 8, Radio Cairo expressed the thoughts of many in the Third World: "The planetary era rings the death knell of colonialism; the American policy of encirclement of the Soviet Union has pitifully failed." Historian Geoffrey C. Ward later recalled "how frightening Sputnik seemed to me as a high-school kid, especially when I got a letter from an old friend in India that simply said 'with this news America is finished,' and asked plaintively 'What happened? How could America let this happen?' as if we had somehow lost control."

Fred Blumenthal, Washington correspondent for the national Sunday supplement *Parade,* decided to see international public opinion for himself and made a tour of Asia in the weeks immediately following the launch.

"Sputnik just turned the East upside-down," he told his readers. "As it did everywhere, I suppose it appealed to the imagination. In the Philippines, the restaurants, movie houses and taxicabs were renamed Sputnik. One restaurant even came up with a Sputnik sandwich, the feature of which was an olive with four protruding toothpick 'antennae.' More important was the diplomatic loss of face [the United States] suffered. In Hong Kong, Japan and the Philippines, there was a loss of confidence in the United States and, under the surface, a sort of secret glee that [it] had been toppled from the high horse."

Opinions about America in the Third World were also influenced by the crisis in Little Rock, Arkansas. The world's press disseminated disturbing and grotesque images of white adults displaying raw hatred toward black children.* America, by its actions, helped the Russians look good to the rest of the world—especially to Third World countries and people of color. The Kremlin used Sputnik as a harbinger of the future—a future marked by world communism—and Little Rock as a reminder of the status quo.†

Another racial incident occurred in the United States less than a week after Sputnik. On October 10, a chagrined and horrified President Eisenhower apologized to Ghana's finance minister, Komla Agbeli Gbdemah, who had been refused service in a Delaware restaurant.

Meanwhile, behind the Iron Curtain, the jubilation was kept in check. The internal boasting was moderate: Sputnik was simply presented to the Russian people as a validation of Marxism-Leninism. Commemorative postage stamps were issued, Sputnik lapel pins worn, poems were published in the papers, and posters proclaimed the achievement a victory over capitalism.

Thirty-five years after the launch, Korolev's biographer, James Harford, interviewed academician Boris Rauschenbakh, who was a friend of

*Chester Bowles, a former ambassador to India and Nepal, looked back at year's end in an essay for *The Britannica Book of the Year* and concluded that Sputnik's impact was magnified because of its proximity to Little Rock. "In newsreels and photographs the colored two-thirds of the world who live in Asia and Africa had seen nine dignified young Negro Americans walking through lines of jeering whites whose faces were distorted with racial hate."

†The Russians used Little Rock as a counterpoint to Sputnik with a degree of subtlety not common to the Cold War. Every day Radio Moscow broadcast in English the moment Sputnik would appear over major U.S. cities and, to put an extra twist on this, over Little Rock.

Korolev's. Harford was told to "look up the pages of *Pravda* for the first day after the launch. It got only a few paragraphs. Then look at the next day's issue, when the Kremlin realized what the world impact was." Harford found the newspapers and reported: "The article in the October 5 *Pravda* was, indeed, tersely phrased. Positioned modestly in a right hand column part way down on the first page, it did not even mention the satellite in its head. Titled routinely, 'Tass Report,' it gave the facts of the launch clinically. . . . The next day's *Pravda* was something else. 'WORLD'S FIRST ARTIFICIAL SATELLITE OF EARTH CREATED IN SOVIET NATION' stretched across the top of page one, which was devoted almost entirely to the achievement."*

*In the days to come, though, *Pravda* was delighted to print the praises of friends and enemies. Interest in the topic of space grew as more Sputniks were launched. The big celebration would not be held until Yuri Gagarin became the first man in space.

CHAPTER SEVEN

Dog Days

Nothing less than *control of the heavens* was at stake. [Sputnik] was Armageddon, the final and decisive battle between good and evil.

—Tom Wolfe, *The Right Stuff*

For postwar America, it was the worst of times—a time of failure, recrimination, and finger-pointing.

Rather than accept Sputnik entirely as a Russian achievement, Americans searched for someone or something in the West to blame. Since nobody in the West knew the identity of whoever had designed the satellite and launch system, the hunt was on to find an accomplice in the West. One rumor, without a shred of evidence, was that it had to have been a British-educated Russian scientist named Peter Kapitza. In an article in the *American Mercury* titled "Britons Helped Win the Satellite Race," Tom Cullen wrote, "The big hero at Cambridge University these days is Dr. Peter Kapitza, the Russian who designed Sputnik and got it off the ground." The logic of the Kapitza argument was that there was no way the Russians could have done this without a major assist from the West in the form of a Western education or something even more insidious.

Sputnik came at a time when American anticommunist sentiment was still a powerful force, and though the bullying and badgering tactics of Joseph McCarthy had ceased, the fear of homegrown subversion through American communism was very much alive, especially among veterans of

World War II and the Korean War. On June 17, 1957, the U.S. Supreme Court had handed down a battery of decisions that, in the words of Patrick Murphy Malin, the director of the American Civil Liberties Union, protected "a freedom of speech and association."*

The *American Legion Magazine* in October 1957 published a rebuke of the decisions: "Not since the U.S. Senate voted 'condemnation' of Senator McCarthy have the communists and their fellow travelers been as ecstatically happy as they are today—thanks to the Supreme Court. They have been joyfully dancing like whirling dervishes in their drawing rooms, their academic halls, their radio and television studios, their Broadway and Hollywood dressing rooms, their union headquarters, their laboratories, and their editorial offices." The mood was right for a post-Sputnik Red hunt.

Spy Story

Throughout the month of October, open season was declared on subversives and spies—even dead ones. Yellowing transcripts of testimony given by informer and convicted spy David Greenglass against Ethel and Julius Rosenberg were brought out to show that American space secrets had been turned over to Russians. First the Rosenberg connection was a rumor. Then it became a front-page story in the *Chicago Tribune* for October 12 under the banner headline "Bare Theft of Moon Data." The story claimed that "Russian spies stole American satellite secrets to enable the launching of the earth 'moon' now racing around the world at an 18,000 mile an hour pace." Evidence of this theft was made available to the *Tribune* "from government sources in the transcript of the trial of Julius and Ethel Rosenberg."

*Typical of the cases was that of *Sweezy vs. New Hampshire,* in which the Supreme Court said that the editor of a Marxist magazine was not required to answer questions put to him during a New Hampshire legislative inquiry into subversive activities within the state. The Court reversed Sweezy's conviction for contempt in refusing to answer questions concerning his "political associations" and about a leftist lecture that he had delivered at the University of New Hampshire, saying in part: "Our form of government is built on the premise that every citizen shall have the right to engage in political expressions and associations. . . . Any interference with the freedom of a Party is simultaneously an interference with the freedom of its adherents."

The next morning, the *Tribune* reported that its source was former U.S. attorney Myles Lane, a Rosenberg prosecutor, who said flat out that Sputnik was the fruit of the "sky platform" data that had been given to the Russians by Julius Rosenberg in 1947. "The fact that the Soviets launched Sputnik last week did not surprise me in the least," he added.*

In response to Lane's charge, the Senate Internal Security Committee announced its plans to hold a full hearing on the matter, and its counsel, Robert Morris, was sent to the Lewisburg Penitentiary in Pennsylvania to take yet another deposition from informant David Greenglass.

It turned out that Greenglass knew no more than he had already reported, and plans for the hearings were abandoned, especially as newspaper editorials suggested that the attempt to blame the Rosenbergs, who had been convicted of espionage in 1951 and executed in 1953, was nothing more than an attempt to divert attention from America's failure. Besides, it was argued, if the Russians had relied on American technology, an American satellite would be in orbit, not a Russian one.

In their book *The Rosenberg File*, generally regarded as the most objective and comprehensive source on the subject, Ronald Radosh and Joyce Milton conclude that Greenglass knew very little about the project, but it was his impression that work on the platform idea had been done somewhere in upstate New York. FBI investigators had forwarded this sketchy description to Air Force intelligence, which readily identified the sky platform as a pioneering proposal for an artificial Earth satellite that had been developed by the RAND Corporation in California shortly after the war.

Edward Scheidt of the New York FBI office reported to J. Edgar Hoover that a comprehensive report on the platform had indeed been in circulation sometime during 1946 or 1947 and might have come to Rosenberg's attention; however, according to Radosh and Milton, "the project was now a dead letter, having been abandoned by the Air Force because, in Scheidt's words, 'the cost estimates were so huge.' "

The report in question was, in fact, RAND's *Preliminary Design of an Experimental World-Circling Spaceship*, its first space study, and it had been distributed to aeronautical laboratories in various parts of the United States, including one in upstate New York—the Aeronautics and Ordnance System Division of General Electric's Schenectady, New York, research facility. This

*According to *Look* magazine, a mathematician named Joel Barr was the one who had turned over information on launching the "sky-platform-earth-satellite."

was the same lab at which a convicted spy who worked with Rosenberg, named Morton Sobell, had been employed as a project engineer until mid-1947. Radosh and Milton point out that the FBI was never able to develop hard evidence that Sobell had actually given the RAND report or segments of it to Julius Rosenberg, so the subject was never mentioned in court. Assuming the full RAND report had made it to the Kremlin, it contained little that Korolov and others in the Soviet Union did not already know.

Ultimately, America rejected the premise that spies had stolen the secret of Sputnik for the simple reason that it seemed to be a secret that had eluded America. "Those Russian spies must be really good," science fiction editor John Campbell said, because "they stole a secret we didn't even have yet."

Did Persecution of U.S. Scientists Impede America's Space Program?

The search for a culprit also went as far in the other direction as to suggest that it was the fear of communism that had put America behind the Communists. For instance, no less a figure than former president Harry Truman suggested strongly that the "persecution" of the best U.S. scientists by Senator Joseph McCarthy during the early 1950s had been a setback for the nation's development of satellites and rockets. Truman's charge was not entirely without merit, as underscored by the case of Hsue Shen Tsien, a brilliant scientist who had served America well—before, during, and after World War II.

Tsien came to the United States on a Boxer Rebellion fellowship and had been on the team that created, among other things, the WAC-Corporal rocket and helped found the Jet Propulsion Laboratory.* He was a protégé of Theodore von Kármán, who considered him his most gifted student, and was selected as one of the first uniformed American officers to interview the Rocket Team in Germany after its surrender. In 1949, just as China was falling to the Communists, Tsien made the decision to become an American citizen.

In 1950 Tsien, now Goddard professor of jet propulsion at Caltech, was

*In her book *The Boxer Rebellion,* Diana Preston notes that after the conflict, arrangements were made to allow young Chinese to study in Europe, Japan, and the United States, "hastening the influx of new ideas into China."

told that his security clearance had been revoked. The FBI believed that a left-leaning political discussion group to which both he and the leader of Caltech rocket development crew at the time had belonged in the mid-1930s was actually a Communist Party cell. While efforts to clear Tsien's name were being made, he attempted to return to China to visit his ailing father and was arrested at the airport. After serving time in jail, he was released but found guilty in deportation hearings. He was then allowed to resume his teaching duties under severe restrictions amounting to house arrest. Five years after his security clearance had been lifted, he was deported in an exchange for American POWs captured during the Korean War who were still held captive in China.

Over the next two decades Tsien became one of the most important scientific figures in China. Historian Michael J. Neufeld wrote that the deportation was probably the "greatest act of stupidity of the entire McCarthyist period . . . The People's Republic now has nuclear missiles capable of hitting the United States, in large part because of Tsien."

It has never been proved whether Tsien was actually a spy or the unlucky victim of Red baiting.* Those who saw the irony of the Tsien story were many. Biographer Iris Chang wrote in 1965: "When the U.S. deported Tsien to China this country deported a man whose American education and training had fully equipped him to transform a backward military power into a nuclear powerhouse capable of delivering warheads to any continent. The legacy of Tsien still haunts the United States. For example, the Silkworm missile menaced American forces during Operation Desert Storm. His deportation sowed the seeds of international nuclear proliferation from Asia into the Middle East. China is now one of the largest arms dealers in the world today, especially to the Third World."

*Iris Chang tried to answer this question in 1995 when her book on Tsien, *Thread of the Silkworm,* was published. "While there exists the remote possibility that Tsien might have been a spy, I believe he was not. The government never provided any substantial proof, either documentary or testimonial, to convince the public that Tsien had been a spy or even a member of the Communist Party in the 1950s. I conducted my own investigation and found that, if anything, Tsien had served the U.S. government faithfully during the war years, receiving official military commendations for his technical achievements. Tsien also exhibited, prior to 1950, a keen desire to stay in the U.S. permanently. He applied for U.S. citizenship and accepted a tenured academic position and arranged for his wife, the aristocratic daughter of a top military strategist for Chiang Kai-shek, to come to the U.S. Even after the INS attempt to deport him, he fought the U.S. government legally through numerous deportation hearings—hardly the actions of a man yearning to live in a Communist country."

Materialism Is Blamed

In the wake of Sputnik, America's self-deprecation was boundless. While some instant experts blamed the schools for not emphasizing science and math, many others charged that the nation's materialistic society was somehow responsible for its loss of technological superiority.

A higher standard of living, seen as prima facie evidence of American pre-Sputnik superiority over Russian communism, now became an emblem of national inferiority. Typical of a number of comments were those of Senator Styles Bridges of New Hampshire, who declared it was now time for Americans "to be less concerned with the depth of the pile on the new broadloom rug or the height of the tail fin on the new car and to be more prepared to shed blood, sweat, and tears if this country and the free world are to survive."

Former congresswoman Clare Boothe Luce said each beep of Sputnik was "an intercontinental outer-space raspberry to a decade of American pretensions that the American way of life was a gilt-edged guarantee of our national superiority." The irrepressible Bernard Baruch noted: "If America ever crashes, it will be in a two-tone convertible."

The following quote gives some indication of how persuasive this anti-materialism argument was. In a letter to stockholders included in the annual report of the National Capital Bank of Washington, Chairman George Didden Jr. wrote: "If Sputnik can't make us realize how unimportant are our desires for color television, can it make us realize the possibility of the loss of all television, except perhaps the few programs monitored and fed to us by Communist zealots?"

Nobody was willing to let a nation of tail-fin lovers off the hook, even a man who used bona fide Texas longhorns for an automobile hood ornament: "It is not very reassuring to be told that next year we'll put an even better satellite in orbit, maybe with chrome trim and automatic windshield wipers. I guess for the first time I've started to realize that this country of mine might not be ahead in everything," said Senate majority leader Lyndon B. Johnson. A *New York Times* editorial chided, "We have become a little too self-satisfied, complacent, and luxury loving." Pundit Walter Lippmann direly noted, "With prosperity acting as a narcotic, with philistinism and McCarthyism rampant, our public life has been increasingly doped without purpose." According to Lippmann, editorialists, and other severe social critics, the nation was adrift in troubled waters.

The Soviets, of course, did not fail to press the point. At an international conference in Barcelona in mid-October, shortly after the Sputnik I launch, Leonid I. Sedov, the prominent Soviet scientist, told an American delegate: "You Americans have a better standard of living than we have. But the American loves his car, his refrigerator, his house. He does not, as the Russians do, love his country."

America's editorial writers had a field day, filling their pages with articles whose theme was the superficiality of much of American technical "progress." In *The Glory and the Dream*, William Manchester concluded, "Sputnik I dealt the *coup de grace* to Ford's fading Edsel, which had been introduced the month before, and which was now widely regarded as a discredited symbol of the tinny baubles America must thrust aside." The Edsel, with its odd oval puckered mouth grille centerpiece was hardly the only example of excess at this time of dazzling grilles, festooned chrome trim, big V-8 engines and three-toned paint jobs. The Cadillac epitomized the "American Dream Car" which was so big and so thirsty for fuel that it averaged only ten miles per gallon.*

The Russians seized this theme to discuss their own progress. "The launching of the Sputnik," proclaimed *Pravda*, "was a victory of Soviet man who, with Bolshevist boldness and clearness of purpose, determination and energy knows how to move forward . . ."

Sputnik II

At the beginning of November 1957, the United States seemed to be starting to return to business as usual. It was a relief when the annoying beeping signal from Sputnik, having gradually weakened, finally stopped; the signal was officially declared dead on October 26 by Radio Moscow, although a silent Sputnik would orbit for another seventy days. There was also a growing hope that the United States might soon be able to launch its own satellite. Vanguard prototype Test Vehicle-2 (TV-2), with dummy second and third stages, successfully met its test objectives on October 23 by reaching an impressive 109-mile altitude and a top speed of 4,250 miles per hour.

*The theme of American excesses would not die and finally went into a grand theatrical finale with the newsreel cameras rolling. The showdown came on July 25, 1959, when Vice President Nixon opened the American National Exhibition in Moscow and tried to impress Khrushchev with America's advances in color television.

On Friday, November 1, the White House announced that the president would soon deliver a series of talks to the American people putting Sputnik, and America's response, into proper perspective. These were to be "confidence" talks patterned after Franklin D. Roosevelt's reassuring Fireside Chats delivered during the Great Depression.

But two days later, thirty days after Sputnik and before Eisenhower's first scheduled chat, the Russians launched Sputnik II. It was not the Moon shot that some had expected but an impressive display of another sort. Sputnik II was the second spacecraft launched into Earth orbit and was the first biological spacecraft. Much larger, at 1,118 pounds (508.3 kg), this second vehicle carried a dog and the life-support system needed to keep the animal alive for 100 hours. The satellite was immediately dubbed "Muttnik."

The first living thing to travel to outer space was a fourteen-pound female mongrel who was part Samoyed terrier. She was originally named Kudryavka (Little Curly) but later renamed Laika (Russian for "barker"). The padded, pressurized cabin on Sputnik II allowed enough room for her to lie down or stand. An air-regeneration system provided oxygen, while food and water were dispensed in a gelatinized form. Laika was fitted with a harness, a bag to collect waste, and electrodes to monitor her vital signs.

Early telemetry indicated Laika was agitated but eating her food. There was no capability of returning a payload safely to Earth in 1957, so—to the horror of many—the Soviets allowed that this was a doomed mission. According to plan, Laika would run out of oxygen and die after about ten days of orbiting the Earth, but because a thermal shield was lost at launching and rendered the thermal control system inoperative, the capsule overheated, killing the dog on the fourth day. However, this was not known—or even suspected—at the time, except by a few Russians working on the program. Sputnik II stayed up for five months, coming down—dead dog and all—on April 14, 1958.

The mission not only provided Soviet scientists with the first data on the behavior of a living organism in the space environment but appeared to cement their dominance of the heavens. American leaders were chagrined that the Soviets seemed almost blasé about Sputnik II, which was much larger and more sophisticated than anything on America's drawing boards. Even the timing suggested Soviet confidence and nonchalance. According to reports and rumors from Russia, the second satellite would go up on the seventh to celebrate the fortieth anniversary of the Bolshevik Revolution, but the Russians had tossed it up four days early. So convinced were American experts that the seventh was the date that 200 Moonwatch teams—

Laika in a hermetically sealed cabin before being placed on Sputnik II. (Author's collection)

IGY-affiliated groups watching for comets, asteroids, satellites, and other celestial events—were alerted to start looking on the fifth. The *New York Times* and the *Washington Post* both said that it would be the seventh. Even after Sputnik II went up on the third, there was speculation that the Russians would still put something else up on the seventh.

The details that the Russians shared in the coming days were impressive. Sputnik II was a four-meter-high cone-shaped capsule with a base diameter of two meters. It contained several compartments for radio transmitters, a telemetry system, a programming unit, a regeneration and temperature control system for the cabin, and scientific instruments. A separate sealed cabin contained the dog. The telemetry system transmitted biological and engineering data to Earth for fifteen minutes of each orbit. Two spectrophotometers were on board for measuring solar radiation (ultraviolet and x-ray emissions) and cosmic rays. A television camera was mounted in the passenger compartment to observe Laika. The camera could transmit 100-line video frames at ten frames per second.

By contrast, the United States was still working to orbit the Vanguard satellite, which weighed a paltry 3.25 pounds. The noted British astronomer Sir Bernard Lovell, upon hearing that the first Sputnik weighed 184 pounds, immediately asserted that a decimal point had been lost and that the weight must be only 18.4 pounds. Sputnik II weighed as much as a four-door full-size Detroit sedan.

"Most of the concern with people like John Stennis and Lyndon Johnson in the Senate and John McCormack in the House had to do with the difference in launch capability between the Soviets and us. The concern was directed at the rockets. Sputnik wasn't a scientific launching. It mostly had to do with rocket propulsion and missiles. Its implications were military, not scientific," recalled Homer E. Newell in a 1981 interview. He was head of the Rocket Sonde Research Branch of the Naval Research Laboratory in 1957 and would later become chief of NASA's Space Science Office.

For John Townsend Jr., Newell's deputy at the time and later president of Fairchild Space and Electronics Company, the biggest surprises following the launch were the extent of U.S. public reaction and the size of the satellite. "None of their early satellites were at all sophisticated," he said. "Instead of building them like a Swiss watch—as we've always done—they'd use common metals, not lightweight alloys. They used radio tubes instead of transistors. But they didn't have to be concerned with sophistication; they had enough lift-power. Later, when I was in Moscow and saw the hardware, it was clear that they had brute-forced everything." Newell added: "If you look at the history of all these early space shots, you find that the Russians picked their missions very carefully. They had two objectives in mind: one, do it carefully; two, make a worldwide impact."

That impact was certainly made. As before, a few American and British scientists suspected the announced weight was an exaggeration. Those who believed it suspected something new. The *New York Times*, for instance, reported that the great stated weight of the satellite "led to speculation that the Soviet Union might be using some new form of rocket propellant unknown in the West."

Several commentators even suggested that Russia might use its missiles to blackmail the United States into a Cold War surrender. One of America's most influential commentators, Edward R. Murrow, reminded America that it could no longer negotiate from a position of strength and that it was time to accept any plan that afforded the opportunity for peaceful coexistence.

Compassion for Laika

If there was any consolation for America in all of this, it was that the planned death of the dog was bad propaganda for the Soviet Union. Concern for Laika became big news. In Britain a minister who refused to pray for the dog encountered protesters and picket lines, while animal rights advocates staged noisy protests in front of the Soviet Embassy in London.

The American Society for the Prevention of Cruelty to Animals said it deplored the use of a dog in an effort that "cannot possibly advance human health and welfare." An animal adoption group in New York termed it "an atrocity" in a telegram to the Soviet Embassy in Washington, and a canine picket line was set up in front of the United Nations. Several of the protesting groups demanded that Laika be brought back immediately, and even American reporters in Moscow had dog-loving Russians innocently asking, "What provision has been made for the dog to come back?"*

With each orbit, the world seemed to take Laika closer to heart. A radio station in Blackfoot, Idaho, claimed to have recorded her barking. The British press gave her all sorts of affectionate nicknames, including "the Hound of Heaven," and a *New York Times* editorial lapsed into pathos and called her the "shaggiest, lonesomest, saddest dog in all history."†

Clearly unprepared for the uproar, the Russians began hinting that the dog might come back by parachute, but this bit of nonsensical disinformation was floated when Laika was dying or dead. In 1965, when Eisenhower wrote his memoir of the period, it still amused him that for many the dog's death overshadowed the stunning feat of getting her up in the first place: "By a strange but compassionate turn, public opinion seemed to resent the sending of a dog to certain death—a resentment that the Soviet propagandists tried to assuage, after its death, by announcing that it had been comfortable to the end."‡

*It has been suggested by one space historian—admittedly without any firm evidence—that the Central Intelligence Agency may have helped stage some of the many protests. CIA Sputnik-Lunik capers, discussed in chapter 8, support this notion.

†Laika also inspired good-natured joking, which helped endear her to many. It was almost as if Walt Disney had put a lovable mutt in orbit. If Sputnik I inspired a few passable one-liners—"The only thing to sphere is sphere itself"; "Around the White House they're calling it the International Geo-Fizzle Year"—Sputnik II and Laika brought out the wags of the world. It was said the Russians were about to send up a Sputnik loaded with cows that would be known as "the herd shot round the world." The dog, we learned, was really an "Airedale," and she put American science deeper in the "doghouse." Skeptics were accused of "barking up the wrong tree," and if the United States would get into the competition it would be a real "dogfight." Aside from Sputnik jokes per se, the other by-product was a sudden rash of Martian jokes in which an alien lands on Earth and, in many of these gags, asks an earthling to "take me to your leader." In one version a Martian approaches Vice President Nixon and asks to be taken to his leader. Nixon replies, "I can't; I hardly know the man."

‡In 1987, close to thirty years after Sputnik II, Laika's death figured in the Swedish film *My Life As a Dog*. The film opens with a soliloquy in which the main character, a young boy named Ingmar, identifies with the hapless Laika, who, as he understood it, was sent up in a Sputnik with wires attached to her heart and brain to see "how she felt" and was then spun around the Earth until she starved to death.

American Speculation Grows

Sending Laika into space also made it clear that the Russians were planning—or at least seriously considering—a manned spaceflight program. The possibility was spoken of glibly as Americans talked about a "man-nik" as the next step. In London outraged British dog lovers were told by an embassy official, "Many humans had volunteered to go into space aboard Sputnik II but a dog was sent instead because the sacrifice [of a human] could not be accepted." This suggested that the spacecraft was ready for humans but not yet tested. Tass, the official Soviet news agency, said its next space passenger might be an "anthropoid ape." Americans, including test pilots like Alan Shepard and Donald K. "Deke" Slayton who began to think in terms of an American manned space program, had no doubts at this point that humans would soon follow.*

Evidence of how "spooked" Americans were, particularly after Sputnik II went into orbit, was the prevalence of very serious talk in the United States that the Russians would set off an atomic or hydrogen bomb on the Moon on the seventh to mark the anniversary of the Bolshevik Revolution; the seventh was to be the night of a lunar eclipse, which would make the explosion much brighter. The chief speculator was Fred L. Whipple, director of the Smithsonian Astrophysical Laboratory, who made it clear that he was theorizing based on indirect and unconfirmed reports attributed to Russian sources. The story had "legs" and appeared in almost every newspaper in the United States and many throughout the rest of the world.

Eisenhower's First "Chins-Up" Talk

On November 7, President Eisenhower went on radio and television to give the first of his "confidence talks." The official White House name for them was "Science and Security," but, unfortunately for Ike, they were quickly dubbed his "chins-up" talk. The name came from a Herblock cartoon in the *Washington Post* titled "Chins Up," which depicted

*Into this menagerie came a report from a game farm operator in the Catskill Mountains of New York to the effect that he had just shipped four man-sized black bears by air to the U.S. Air Force Missile Development Command in New Mexico. The Air Force did not deny the purchase nor say why it had purchased the animals, but insisted that the bears would not be sent into space.

During his November 7, 1957, televised address to the nation President Eisenhower displays the recovered Jupiter C nose cone, which was the first man-made object recovered from space by the United States. (*U.S. News & World Report* Collection, the Library of Congress)

a frowning Ike with "Confidence Speech #1" on his desk while he looked chin-up at an object in the sky marked Sputnik II.

The big news in that first speech was the appointment of James R. Killian Jr., president of the Massachusetts Institute of Technology, as the first White House science adviser. "According to my scientific friends, one of our greatest and most glaring deficiencies is the failure of us in this country to give high enough priority to scientific education and to the place of science in our national life," Eisenhower declared. He also promised to put an end to the inter-service rivalry in the Pentagon, which appeared to be holding back the American missile program. Sitting at his desk, the president pointed to a large conical object on the floor to his left, a missile nose cone recovered from suborbital flight on a Jupiter C rocket a few days before—the first object ever recovered from space. With two Sputniks in orbit, this display of a suborbital souvenir at the president's feet in the Oval Office was hardly reassuring.*

*Although miffed that the president had not given them credit for the launch and recovery of the nose cone, the Huntsville contingent was soon using this accomplishment to their advantage. Medaris told an audience in Detroit on November 18 that it indicated that the United States might be able to send a man into space and bring him back alive.

Late that evening, Eric Sevareid summarized the day's news on the CBS radio network from Washington: "All day today there has been much speculation here that two kinds of surprises might be forthcoming tonight; one from the Russians—a shot at the moon, perhaps—and something from the President in the nature of some hitherto unrevealed American rocket achievement of dramatic consequence. Neither event transpired."

Despite an outward appearance of calm, which seemed to border on apathy, the Eisenhower administration was deeply upset by the success of the Sputniks and determined to respond in kind and in reasonable time.

Send in the Rocket Team

The day after Eisenhower's first speech, thirty-five days after the first Sputnik, and five days after Sputnik II, the Department of Defense issued the following press release, giving the go-ahead for Medaris and von Braun to launch the Jupiter C as soon as possible.

> *The Secretary of Defense today directed the Department of the Army to proceed with launching an earth satellite using a modified Jupiter C. This program will supplement the Vanguard project to place an earth satellite into orbit around the earth in connection with IGY. All test firings of Vanguard have met with success, and there is every reason to believe Vanguard will meet its schedule to launch later this year a fully instrumented scientific satellite. The decision to proceed with the additional program was made to provide a second means of putting into orbit, as part of the IGY program, a satellite which will carry audio transmitters compatible with Minitrack ground stations and scientific instruments selected by the National Academy of Sciences.*

However, the official directive sent to Huntsville differed from the press release in one important regard. It did not say "to proceed with launching" but rather "to prepare to launch." Medaris was furious and shot off a telegram to the secretary of the Army and General Gavin in which he flatly stated that this was unjust and unacceptable, and that if his group could not have firm permission to go ahead and fire a satellite into orbit, they wanted to quit and move on to other work. Von Braun and William H. Pickering, the director of the Army's Jet Propulsion Laboratory (JPL) who was in Huntsville at that moment, were ready to quit. Pickering had just joined the team and brought with him an actual satellite and the scientific experiments to ride with it.

Unknown to Eisenhower and almost everyone else in Washington, including the secretary of defense, the Huntsville crew had been preparing to launch since October 5—a crash effort without sanction or budget. Despite threatening to quit if they did not get their permission worded exactly the way they wanted, the crew were now America's best hope to get into orbit and they knew it.

They were also playing with fire. Or at least they thought they were. One of Eisenhower's last orders to Charles E. Wilson, his retiring secretary of defense, was a directive on the morning of October 8 "to have the Army prepare its Redstone at once as a backup for the Navy Vanguard." For a reason or reasons obscured by time, the permission was never sent, and the official OK did not get to Huntsville for a month. What the folks at ABMA and JPL should have been afraid of was the discovery that they were working covertly and in collusion, preparing for a satellite shot years before getting authorization to do so. This had long bothered the Vanguard team, which suspected as much. Constance McLaughlin Green and Milton Lomask reported in their history of Vanguard that when, in a November 1966 interview, they apprised former president Eisenhower of this frequently advanced charge of unauthorized Army satellite work, he expressed surprise, saying "that would have been a court martial offense!"

Medaris and company were finally assured that the language of the permission had no intended restriction or hidden meaning, and that they would be given the chance to launch two satellites early in the new year. However, this would be under the condition that they stopped embarrassing the president by talking out of turn. General Lyman Lemnitzer, Army chief of staff, told Medaris on November 9 that von Braun had given an interview on the West Coast that was so damaging to Eisenhower's effort that the head of the Associated Press had to be persuaded to censor the more incendiary comments. Medaris—an authoritarian, old-school general who affected a swagger stick—then timidly, and unconvincingly, suggested that he could not exercise "complete censorship" over the von Braun team.

Medaris was told in no uncertain terms that "the time for talking has stopped" and to bring himself and his Germans under control. Period. With the semantic issue resolved and a gag on presidential attacks in place, America could now position itself for recovery. On November 14 Medaris and von Braun came to Washington to brief defense officials and the press on their plans. Medaris told a packed press conference that there was a 90 percent chance that the Army would put a twenty to twenty-one-pound, bullet-shaped satellite into orbit on the first attempt.

From this moment forward the Army would give the impression that it was just beginning to work on a launch vehicle and satellite. "Redstone Starts Building Satellite" was the headline on a UPI story in the next day's *Birmingham Post-Herald* (Alabama). Muckraking columnist Drew Pearson, however, reported that there were "satellites" resting in a warehouse in Huntsville. Medaris emphatically denied this but added that there were some missiles left over from the Orbiter project that was canceled in 1955. He did not allow that the satellites Pearson wrote about were, in fact, real and stored in a locker in Pasadena at JPL. At one point in his November 14 press conference, he was asked point-blank if work had already begun on the Army satellite and admitted that it had, but left the impression that it had just started.

Eisenhower on the Ropes

On November 14, Eisenhower flew to Oklahoma City to deliver his second confidence speech. This time the subject was deterrence and defense. He asserted that he would not sacrifice security to worship a balanced budget and that all that America wanted was adequate security—nothing more and nothing less.

In addition, Eisenhower promised that the United States would now hasten the process of building more strategic long-range bombers and bases on which to place them, while developing long-range missiles.

Eisenhower was embracing bombers in public at a time when the Soviet Union was betting all of its chips on missiles as the weapon of the future. A poker game was being played.

At least a year earlier, Russia had implied that it was going to build a massive fleet of manned bombers to overpower and outnumber the Strategic Air Command. The Soviets had built a bomber—known as the Bison—able to carry a nuclear warhead all the way to the United States. The problem was that America did not know if there was just one prototype or if hundreds of these planes had been built. The Air Force and the defense contractors were eager to start building a U.S. fleet of long-range nuclear bombers.

But Eisenhower refused to accept rumors and raw speculation of a massive fleet of bombers in the Soviet Union. On July 4, 1956, the super-secret U-2 spy plane was sent on its first mission over the Soviet Union to find out the truth. The U-2 discovered, to Eisenhower's satisfaction, that there was no bomber gap.

In the days following the launch of Sputnik II, Nikita Khrushchev was calling the long-range bomber an anachronism, signaling his plan to move right into the age of intercontinental ballistic missiles. On October 8 Khrushchev had met with two members of the British Parliament who were visiting Moscow and told them, "The age of the bomber is over—you might as well throw them on the fire." He added that sending flesh and blood against missiles made no sense anymore.

In Oklahoma City Eisenhower responded by saying: "Today, as I have said, a principal deterrent to war is the retaliatory nuclear power of our Strategic Air Command and our Navy. We are adding missile power to these arms and to the Army as rapidly as possible. . . . But it would be some time before either we or the Soviet forces will have long-range missile capacity equal to even a small fraction of the total destructive power of our bomber force."

Two days after the Oklahoma City speech, Secretary of Defense McElroy announced that steps would be taken to speed intermediate-range missiles to Europe. In the weeks following the Sputnik II launch, the United States pulled out all the stops to help regain the nation's confidence. Some of these demonstrations were impressive—Thor and Jupiter IRBM firings to a height of 2,700 miles—whereas others were pure puffery, such as a nonstop flight to Argentina and back by Air Force general Curtis E. LeMay in a new KC-135 jet tanker.

Of all the demonstrations, the most impressive was the Air Force's launch of an Aerobee rocket on October 16. It proved that the United States was getting close to the dual capacity of intercontinental missile power and the ability to go into deep space. The rocket rose from the New Mexico desert to a height of thirty-five miles, at which point it dropped away and the nose section then rose to fifty-four miles, where shaped charges (a charge shaped to concentrate its explosive force in a particular direction) sent pellets flying beyond the Earth's atmosphere at a speed of 33,000 miles per hour. An extraordinary feature of this experiment, which was not made public for a month, was that, as the speed of the pellets was greater (by 8,000 mph) than the velocity needed for an object to escape the Earth's gravitational pull, they continued to travel through interplanetary space with nothing to stop them.

Such displays notwithstanding, Eisenhower suffered in the days following Sputnik II, when his Gallup poll rating dropped twenty-two percentage points, and the criticism of his stewardship redoubled.

Just when it seemed that things couldn't get any worse, they did. On November 25, as President Eisenhower was getting ready to deliver his third "confidence" speech, he was enjoying a light lunch while seated at his desk. He suddenly felt dizzy and was unable to hold on to a letter in his hand. When he did eventually pick up the letter, the words on it appeared to run off the top of the page into thin air. President Eisenhower had suffered a stroke. When he tried to tell his secretary, Ann C. Whitman, what had happened, his words came out as gibberish. Later in the day, he regained enough speech to tell Mamie Eisenhower that he feared his days in office had come to an end.*

The second blow was the hearings before the Senate Preparedness Subcommittee under the direction of Lyndon Baines Johnson. In his opening remarks on November 25, he not only compared the Sputnik crisis to Pearl Harbor but suggested that Sputnik was an even greater challenge. "In my opinion," he said, "we do not have as much time as we had after Pearl Harbor." Despite that opinion, Johnson took the rest of November, all of December, and most of January to conduct the hearings, bringing in a steady stream of seventy-eight distinguished witnesses who generated 2,313 pages of testimony.

The serious effects of Eisenhower's stroke were short-lived, although as the third serious illness of his presidency it seemed to sap his vigor. The effects of the hearings were, in a sense, worse in that they focused on the many shortcomings of America in missile development, space, education, science, and more.

The politics of the Johnson hearings were brutally simple and partisan, although they were successfully staged as bipartisan. Washington understood that the hearings were political. "LBJ was eager to get out front in space because it was the new national toy. He was trying to get to become President of the United States," recalled White House insider Bryce N.

*Richard Nixon, John Foster Dulles, and others close to the president speculated that the stroke was caused by the Sputnik criticism. Eisenhower himself saw it as due to a prolonged period of stress. A week before the sudden illness he wrote to a friend, "Since July 25th of 1956, when [Egyptian president Gamal Abdel] Nasser announced the nationalization of the Suez, I cannot remember a day that has not brought its major or minor crisis. Crisis had now become 'normalcy.' "

Senate Preparedness Subcommittee Chairman Lyndon B. Johnson examines a model of the Vanguard satellite on November 27, 1957, during a recess from the hearings on rockets, missiles, and space. (*U.S. News & World Report* Collection, the Library of Congress)

Harlow in a June 1974 interview with NASA historian Eugene M. Emme. Harlow saw Johnson's posture as politics as usual: "That's the way you do these things as leading politicians. . . . You do like . . . Robespierre in the French Revolution. There goes the crowd, I must get in front of them, I'm their leader. Okay, that's the way they do this. And so LBJ wanted to get in front of the space rush so that everybody would say, 'Oh, that's our leader.' "

Many themes played out during the next two months as the subcommittee heard testimony from seventeen expert witnesses across the full spectrum of scientific, technological, industrial, and military life. The first theme was that America was in deep trouble. At least two witnesses said that it was going to take as much as ten years for the nation to catch up to the Soviet Union in space.*

Several witnesses considered Sputnik to be the most important military

*On the rare occasions when a witness said that the situation was not dire and that there was no reason to panic, Johnson reacted with homespun cynicism. Donald Quarles, for example, testified that "taking the missile program as a whole and comparing [the Soviet] program with our own, I estimate that as of today our program is ahead," and that for this reason there had been no acceleration of the U.S. rocket program since Sputnik I. Johnson interpreted this as complacency on the part of the Pentagon and the Eisenhower White House. "The net of it is," Johnson said with a drawl, "that the American people can have adequate defense and eat their cake too; and even have whipped cream on it."

development of the post–World War II era. When asked whether the Russians could put a "hydrogen warhead on the city of Washington," Wernher von Braun replied: "I would think so. Yes, sir." Physicist Edward Teller testified that "[Sputnik] has great military significance because, among other things, it shows that the Russians are far along, very far along, in rocket development."

The second theme of the hearings was that the Eisenhower White House and top officials in the Department of Defense had squandered America's advantage in space. Vanguard director John P. Hagen told Johnson that Project Vanguard could have beaten Sputnik into orbit if it had been afforded a higher priority. He reported that he had asked for a higher priority in 1955 but never received a response.*

The real proof of an advantage squandered lay with the Army's repeated point that the September 20, 1956, Jupiter C launch should have contained a satellite. Physicist James A. Van Allen chimed in with a written statement asserting that the Army could have placed a fifteen-pound satellite in orbit. Although he had worked on the Vanguard program until the end of October, he regarded the decision to go with Vanguard as "thoroughly ill advised."

Army generals joined the attack on the administration. General James M. Gavin, for example, said that the Huntsville team was turned down five times and that he, Medaris, and von Braun were so concerned that the Soviets were going to go into space first that they talked about the possibility of shooting down the first Russian craft ("denying satellite intrusion" was the phrase he used) in the event that it was "a military satellite rather than a small, let us say, purely scientific satellite."

The clear implication was that the president and the secretary of defense had hurt the country by not listening to and submitting to the leadership of Gavin, Medaris, and von Braun.

The third theme of the Johnson hearings was introduced by the scientists, engineers, and academicians called in to testify. Instead of fearing or embracing rockets as weapons, this group saw them as vehicles that could produce benefits for humankind. Their testimony created fewer headlines and lacked the charismatic appeal of the Gavin, Medaris, and von Braun

*In the middle of Hagen's testimony Johnson broke in to announce that President Eisenhower had suffered "another form of heart attack." But the single-minded Johnson was not about to adjourn to find out how the president was faring. At another point in Hagen's testimony a lightbulb from an overhead chandelier fell on the committee table. Hagen was asked by Johnson if it was part of his project. He replied: "No sir; I think that it is one of those strange flying objects."

testimony, but these experts made their point effectively. With few exceptions they wanted space taken away from the military and put into a new civilian space agency. Historian Eilene Galloway at the 1997 conference "Reconsidering Sputnik" remarked that for this civilian scientific and technical community, "the problem that required solution was two dimensional: to preserve outer space for peaceful exploration and uses, and prevent its becoming a new arena for warfare. This situation was a classic case of presenting a choice between good and evil."

Launchpad Debacle

When the first Sputnik was launched on October 4, 1957, much of the promise of the future seemed to shift to the East. Then came Sputnik II on November 3; the world now looked to America for its response, which would have to be Vanguard.

Because of technical problems, the first test launch (Test Vehicle-0) for Vanguard did not occur until December 8, 1956, followed by the second (TV-1) on May 1, 1957, and the third (TV-2) on October 23, 1957. Just prior to Sputnik I, the team at the Naval Research Laboratory was working hard to get the Vanguard satellite into orbit on a Viking rocket, but they labored with trepidation and a sense of scientific sobriety. The night that Sputnik was launched, Walter Sullivan reported in the *Times* that the Vanguard team at the Soviet Embassy party breathed a collective sigh of relief after their initial disappointment that they would now have to be second. Vanguard director John P. Hagen told Sullivan, "Now we can concentrate on doing a good job."

During the week of October 7, Hagen was telling President Eisenhower that TV-3, the fourth in the series and the first fully operational rocket with all three of its stages "live," would be strictly experimental. The Vanguard satellite would rest atop the rocket and an attempt to reach orbit would be made, but any public announcement would be premature and misleading. The official satellite launch was due to be made on the sixth rocket (TV-5) of the series.

Despite this caution by Vanguard's manager, a bold White House decision was made to attempt a public satellite launch aboard the next available Vanguard rocket, TV-3. To Hagen's chagrin and horror, presidential press secretary James Hagerty, in a White House press release on Friday, October 11, proclaimed that the December Vanguard would launch a satellite-

John P. Hagen, director of Project Vanguard, addressing the National Press Club on October 16, 1957, outlining America's response to Sputnik. He is shown with full-scale models of two unproven Vanguard satellites. The smaller satellite would become Vanguard I, while the larger, fully instrumented model in the foreground would orbit as Vanguard II, carrying instrumentation to measure the global distribution and movement of cloud cover. (*U.S. News & World Report* Collection, the Library of Congress)

bearing test vehicle. This was universally read as a promise to put America's first satellite into orbit and was taken by Hagen and his team as a command to put a satellite in orbit with Vanguard, a brand-new and untested rocket.

As the time for the Vanguard launch drew near, there was at least one bad omen: The national trade union representing the 600–700 clerks at Cape Canaveral threatened to strike and throw up picket lines. However, the strike was averted, and a great feeling of exaggerated optimism now took hold among the public and the press.

The big day arrived two months to the day after the Russian launch, on Wednesday, December 4, 1957. Reporters from around the world made a mad dash for press credentials, and thousands of Americans—"bird watchers"—flocked to the beaches near Cape Canaveral to watch from deck chairs as America attempted to gain control of its destiny and regain lost prestige.

Because of high winds and assorted "bugs," the scheduled countdown for the Vanguard launch had to be rescheduled for the next day, December 5, when, coincidentally, the 347th meeting of the National Security Council was held. The announcement and attendant humiliation of the previous day's public postponement took up the last minutes of the regular Thursday meeting.

Secretary of State John Foster Dulles, for one, was furious about the circumstances of the delay at the cape and said that he hoped in the future the United States would not announce "the date, the hour, and indeed the minute" of a launch. This information should not go out until the satellite was successfully in orbit, declared Dulles, adding that the effect of the publicity of the last few days, culminating in the final decision to postpone the attempt to launch America's first Earth satellite, had a "terrible effect" on the foreign relations of the United States.

Dulles was concerned with more than mere impressions. He was an architect of the American policy of deterrence and had told the world in 1954 that Communist attacks would be met with massive retaliation by American nuclear forces. Deterrence would be maintained by having "a great capacity to retaliate, instantly, by means and at places of our own choosing." The Sputniks and now this very public delay brought this ability to retaliate into question.

The president commented that he was all for stopping such unfortunate publicity, but he had no idea how knowledge of delays could be stopped in an open society.

The scheduled launch was postponed due to a frozen valve, a fatigued launch crew, and strong winds that threatened to blow the vehicle off course. Finally, at precisely 11:44:55 A.M. on Friday, December 6, 1957, with the whole world watching, the slender vehicle rose a few feet off the launch platform, shuddered slightly, buckled under its own weight, burst into flames, and collapsed.

Kurt Stehling, who headed the Vanguard propulsion group, offered an eyewitness account: "It seemed as if the gates of hell had opened up. Brilliant stiletto flames shot out from the side of the rocket near the engine. The vehicle agonizingly hesitated for a moment, quivered again, and in front of our unbelieving, shocked eyes, began to topple. It sank, like a great flaming sword into its scabbard, down into the blast tube. It toppled slowly, breaking apart, hitting part of the test guard and ground with a tremendous roar that could be felt and heard even behind the two-foot concrete wall of the blockhouse and the six-inch bulletproof glass. For a moment or two there was complete disbelief. I could see it in the faces. I could feel it myself. This just couldn't be."*

The Cape Canaveral "launch" had lasted a mere two seconds before

*The blockhouse was about 300 feet from the launch site. Moments after the explosion the emotion turned from disappointment to fear of toxic gases. "Propellants got into the cable trench and fumes were coming into the blockhouse," launch director Bob Grey recalled many years later. Grey was quoted in Milt Salamon's *Florida Today* column of March 4, 1998. Salamon's enduring fascination with the individuals and events that marked the early days of space exploration make his column must-reading for anyone reconstructing those times.

Remains of the Vanguard TV-3 explosion, the thrust of which sent the satellite itself into the tall weeds near the launchpad on December 6, 1957. (KSC Archive)

turning into a spectacular fireball. Its tiny 3.2-pound payload, thrown free of the fire, rolled into the scrub brush and started beeping. A group of reporters watching through binoculars from the balcony of the new Vanguard Motel were horrified.

A fire control technician flooded the wreckage with thousands of gallons of water. Minutes later the air was cleared, and the project's field manager issued the order, "OK, clean up; let's get the next rocket ready."

By the end of the day, this ignominious failure was being declared a "national disaster." Senator Lyndon B. Johnson spoke for millions when he termed this public display of failure "most humiliating." Within days, television and newsreel film replayed the liftoff and explosion throughout the world.

The reaction to the coupled events of Sputnik II and the explosion of Vanguard was spectacular and intense. What the press, politicians, and public had not known was that the Vanguard team knew they were dealing with a new and untested rocket, and so failure at this stage was not unexpected. Largely forgotten was the fact that the three-stage rocket was designated TV-3, for Test Vehicle 3, and it had been originally scheduled to be just that, a test. Pressure from the White House had prompted turning the test into a full-fledged attempt at a satellite launch several months ahead of schedule.

Three days prior to the explosion, Martin Trimble, a vice president of the Glenn L. Martin Company, had flatly predicted at a business meeting in Melbourne, Florida, that the rocket would *not* put a satellite into orbit. Trimble's assessment was based on the "prevailing mathematics of trial and error"—three failures in every seven tries—"in this kind of testing experiment." At a news conference in Chicago that same day, the chairman of the IGY committee, Joseph Kaplan, was only a tad more optimistic. Cautioning reporters about "risk of failure," he assured them that before the end of the International Geophysical Year on December 31, 1958, the United States "will have a full-fledged earth satellite in orbit."

Following the explosion, Vanguard's small corps of defenders were overwhelmed by legions of detractors. At a press conference in Washington, Hagen chided the newspapers for giving their readers unreasonable expectations of success. Murray Snyder, the assistant secretary of defense for public affairs, argued that there was no excuse for the "exaggerated optimism" with which reporters had covered events leading up to the unfortunate explosion. "The Department of Defense," he said, "exercised great restraint in its announcements, stressing the fact that a preliminary test was involved and that if the test satellite was successfully put into an orbit that would exceed the purpose of the test."

But the nation needed a whipping post and got it with Vanguard. The *New York Times* ran a mocking editorial dubbed "Sputternik." The *New York Herald-Tribune*, one of the administration's staunchest backers, called it "Goofnik," and *Time* magazine smugly suggested that there were many who thought that it should be renamed "Project Rearguard."

Soviet premier Nikita Khrushchev derisively called the tiny Earth-bound Vanguard satellite "an orange." He had already found irony in the Vanguard name before the debacle. In a November 6 speech made to celebrate the fortieth anniversary of the Russian Revolution, he said, "It appears that the name Vanguard reflected the confidence of the Americans that their satellite would be the first in the world. But . . . it was the Soviet satellites which proved to be ahead, to be in the vanguard."

Behind the Iron Curtain, there was no need to go beyond the simple device of repeating acid quotes and devastating Vanguard headlines from U.S. and European newspapers. *Pravda* reproduced the front page of the *London Daily Herald*, which showed a photo of the Vanguard being readied on the launchpad next to the one of the explosion. Superimposed above the immense headline, which read, "Oh, What a Flopnik!" was *Pravda*'s comment, "Reklama and Deistvitelnost," or "Publicity and Reality." Headlines carried by newspapers in Britain and Canada read: "Ike's Sputnik Is Dudnik," "U.S. Calls It Kaputnik," "Ike's Phutnik!" The BBC invited a calypso singer, Rory McEwen, to perform a special song he had written in honor of the failed satellite, which contained these lines:

They've been pressing the button for a month or more,
But they can't get the blighter off the floor.

On the morning after the explosion, sell orders on stock in the Martin Company, the prime contractor for Vanguard, reached such proportions that at 11:50 A.M. the governors of the New York Stock Exchange suspended trading in it—a rare financial event in those days.

But this was just the beginning. On the same morning, at the United Nations headquarters in New York City, the Soviet delegation formally offered financial aid to the United States as part of a program of technical assistance to backward nations. America's humiliation was complete.

The Aftermath of the Vanguard Failure

Following the Vanguard failure, the Johnson hearings were running at full speed, and as the parade of witnesses continued, each was more devastating to the Eisenhower administration than the one before. Halfway through the hearings, on December 16, Johnson issued a press release stating that America's problem with its rocket and missile programs was not with men like Hagen but with those at the top, who could not make "hard, firm decisions." He said that he was speaking for the American people, who in a Gallup poll released the day before laid most of the blame for the nation's missile gap on Eisenhower and the Republicans. In second place was inter-service rivalry.

Weaknesses in American society, first noted after Sputnik I, now came under even more serious fire. These included: insufficient support of science, the sorry state of American education and especially of science and foreign-language instruction, bickering between the military services on technical matters, and even government red tape.

Quickly the message delivered by the press went from the assertion that the American way of life was lacking to the idea that the USSR was forging a better one. The *New York Times* weighed in with a series of articles concerning the strength of the Soviet educational system, with Moscow correspondent Benjamin Fine reporting as early as November 1957 that the Soviet Union was far outstripping the United States in its emphasis on technical and scientific education. For every American weakness, there seemed to be a newly discovered Russian strength. For example, the United States was turning out 22,000 engineering graduates a year; the Soviet Union, 66,000 a year.

Senator William F. Knowland (California) was one of a few critics who realized that since the Soviet program was not totally public like the American program, the world had no idea whether it experienced setbacks similar to those with Vanguard. He pointed out that "the Soviet Union may well have had a dozen before they launched the first Sputnik." They did not, but his comment was prophetic. In 1958 the first attempt to launch Sputnik III would fail, but would be kept secret for years.

Despite the sometimes hysterical reactions, Americans had reason to worry. As Colonel-General Vladimir L. Ivanov, commander of the Military Space Forces, said in an interview released on October 4, 1996, as part of an Official Kremlin International News Broadcast: "The first Sputnik, the automatic probes to the Moon, the historic flight of the manned Vostok with Yuri Gagarin on board—all this was accomplished on the basis of the 'Seven,' the R-7 intercontinental ballistic missile that was unique in terms of design and technological execution. These and all subsequent launches were performed by military missile test personnel."

He then went out of his way to stress the military importance of space in the Soviet era, which was fast coming to a close: "Outer space offers unique opportunities for intelligence, communications, troop control, navigation, cartography, for collecting weather data and fulfilling other important tasks both on a global scale and in the real time mode. In other words, the use of space systems makes it possible to increase by 50–100 percent the efficacy of weapons and military hardware . . ." October 4, 1996, was not only the anniversary of Sputnik but also Military Space Forces Day.

A Bad Report Card

Fourteen days after the Vanguard debacle came the year's final indignity for the United States: The public was informed about a devastatingly frank, top-secret report. In early 1957, the Eisenhower administration had created a blue-ribbon commission to review the state of U.S. civil defense in the event of a Soviet nuclear attack. On its own, the commission broadened its mandate to include the capability of the U.S. military, particularly Strategic Air Command's (SAC) forces, to survive a nuclear attack and still carry out their retaliatory mission. Eisenhower was given the top-secret report as two briefings, one on the fourth of November—the day after Sputnik II—and the other on the seventh.

Called the Gaither report in recognition of the commission's first chairman, H. Rowan Gaither Jr., the forty-page report, which had been turned over to Eisenhower's staff on November 7, said America was most vulnerable and warned that it would be unable to defend itself against a Soviet attack. For starters, there seemed little likelihood of SAC's bombers surviving a missile attack since there was no way to detect such an attack until the first warhead landed. It emphasized both the inadequacy of U.S. defense measures designed to protect the civilian population in the event of nuclear attack and the vulnerability of the nation's strategic nuclear forces. It called for a missile system to defend the country and a massive drive to build fallout shelters to protect Americans from nuclear fallout. It also recommended an $8-billion increase in annual defense spending for a speeded-up missile program, for dispersal and protection of the nation's bombers, and for greatly expanded scientific research and development.

While Eisenhower and his top advisers were still mulling it over, the substance of the report was related to Chalmers Roberts of the *Washington Post* by several members of the committee. On December 20, 1957, the *Post* ran a two-column story on the front page under the headline "Secret Report Sees U.S. in Grave Peril." This story of the Gaither report was reprinted widely, causing the American public to become, to use the words of Eisenhower in his memoirs, "bewildered and upset."

Many people, including Lyndon Johnson and the Democratic Party, demanded that the full text of the report be released. The White House refused, claiming executive privilege but inadvertently giving the impression that Eisenhower had something to hide. Ironically, subsequent presidents, including John F. Kennedy and Lyndon B. Johnson, refused to declassify the Gaither report. In the early 1970s, in the wake of the release of the Pentagon papers, the *New York Times* petitioned Nixon's National Security Council for the report, but the request was turned down. An appeal to the Interagency Classification Review Committee, chaired by Ike's son, John, overruled the council, and the report became public in early 1973.

Why was the text of the Gaither report such a closely guarded secret for fifteen years? It was incredibly pessimistic, even alarmist—the last thing America needed in the fall of 1957 or during the rest of the 1950s because it warned of significant ICBM delivery capacity by 1959. It created the idea of the "missile gap." It also relied on the most secret data in government files, including top-secret U-2 reconnaissance photographs of the Soviet Union. The rules of intelligence and the rules of poker are the same: Never tip your hand. To read the actual report was to get a good glimpse of the state of the art of American espionage and reconnaissance. In an oblique reference to the U-2 flights, which had begun a year prior to the writing of the report, and spy satellites, which were still on the drawing boards, it urged "exploitation of all means presently at our disposal to obtain both strategic warning and hard intelligence even if some risks have to be taken." Ike's secret plan to gather intelligence from the air and then space was validated. Presumably, the last thing that he wanted was to turn this into a public debate, replete with congressional hearings. Finally, there was a sense of fatalism about the report, which came close to saying that America was losing strength that it would never regain. The report suggested that *"this could be the best time to negotiate from strength, since the U.S. military position vis-a-vis Russia might never be as strong again"* (italics in original).

Nuclear Fears Escalate in the United States

The 1950s in the United States are often portrayed as a period of relative calm, quiet, and complacency.* Yet nothing could be farther from the truth. People had good reasons to feel threatened and fearful. Hollywood reflected those anxieties and fueled them by producing a steady stream of movies based on general feelings of lost confidence, confusion, and horror. For instance, in the 1954 film *Them!* giant ants, inflated by a blast of nuclear radiation, come out of the sewers of Los Angeles and begin eating people alive. Even before the Sputniks, Vanguard, and the Gaither report, America was, in the words of Charles C. Alexander in his book *Hold-*

*To the generation of Americans who had been born during the Great Depression and fought in World War II and Korea, the recession year ended badly, right down to "that damn music" their kids were playing. In late 1957, the influence of rock 'n' roll was ubiquitous, even intruding on Christmas. "Jingle-Bell Rock" reached the top of the Christmas-music chart that year—a whole century after the uptempo "Jingle Bells" was written by James Pierpont—and to many older Americans, it was a sound as discordant and depressing as the beep of Sputnik.

ing the Line, in a "collective mood of deep anxiety, often bordering on hysteria" in the summer and early fall of 1957.

Nineteen fifty-seven ended on a decidedly sour note. The rhythm of American life had been interrupted, and odd themes were playing out. Not only was nuclear war a possibility, it grew more likely as the world stocked up on atomic and hydrogen bombs. In the very same issue of the *Post* that carried details on the Gaither report, a story from Paris reported that the North Atlantic Treaty Organization (NATO) had agreed to arm its members with nuclear weapons against the growing Soviet power.*

After the Sputniks and the leak of the Gaither report, millions of Americans seriously considered digging fallout shelters in their own backyards. Civil defense administrator Leo A. Hoegh told city managers to "bring their civil defense plans out of mothballs and bring them up with the times." The assumption was that if $25 billion were spent on fallout shelters, the program could save 50 million people, a reduction of 35 percent in casualties in the event of nuclear war.

Eisenhower rejected the Gaither fallout proposal. At a National Security Council meeting on January 16, 1958, he said, "Talking about such figures, we were talking about the complete destruction of the United States." Vice President Nixon said he did not believe the United States could survive such an all-out nuclear war. "Our major objective must be to avoid the destruction of our society," he declared.

If Sputnik and the leaked Gaither report got the blame for the rise in the purchase of fallout shelters, little unsettling things—such as the strange business of the Russians charging that parts of the Sputnik I carrier rocket had fallen on American soil—added to the national feelings of anxiety and foreboding. Nikita Khrushchev said at a party at the Finnish Embassy in Moscow on December 6 that he knew the Sputnik rocket debris landed in America and was sure that the biggest piece had been found, but that American authorities were refusing to give it back. The next day, the Soviet Academy of Sciences issued a statement saying that the pieces landed in Alaska and on the U.S. West Coast. The debate went on for months and ended with an American scientist, Robert Jastrow of the Naval Research Laboratory in Washington, using orbital calculations to map Sputnik's movements

*The late Mike Royko of the *Chicago Tribune* argued in print that the reason we think of the 1950s as our last tranquil decade was because of the demeanor of Eisenhower. ("He was the last president who respected the right of all Americans not to believe that the sun, the moon and the stars rotated around the occupants of the White House.")

to show that the carrier rocket actually landed in the Gobi Desert in Soviet-controlled Mongolia. What was unsettling about this was not the debate over chunks of scrap metal but the implication that chunks of scrap metal would be falling all over the place as the Space Age progressed.

The sense of eeriness was compounded by a prank involving Sputnik II's radio. The radio signal supposedly died in early December. But on Christmas Eve of 1957, two weeks after the radio faded out, a strong signal reappeared in the California mountains. The Federal Communications Commission was alerted, and two cars driven by engineers and equipped with radio direction finders were sent out to look for its source. The fear was that the radio might have been deposited there as an act of war to serve as a beacon that would guide Soviet missiles into Los Angeles. After a forty-two-hour hunt, the engineers found the radio in a cactus-covered canyon on Mount Wilson. It turned out to be a joke—four aircraft engineers had placed it there in the hope that it would be mistaken for the satellite—but it didn't get many laughs in a jittery America.

The pranksters were convicted and fined.

The Great Flying Saucer Revival

With so many people looking skyward in 1957, it was probably inevitable that a great Sputnik-inspired UFO revival was forthcoming. The term *flying saucer* had come into the language a decade earlier when Kenneth Arnold of Boise, Idaho, reported seeing flying saucers over Mount Rainier, Washington. It was the first of many such reports of "shining saucerlike objects." In 1952 there had been a near panic over them, but few notable sightings had been made for some time. On October 25, 1955, the Air Force had released the conclusion of its top-secret Project Saucer: that many reports of flying saucers were merely illusions or explainable as misinterpretations of "conventional phenomena." By the end of 1955, the average citizen had all but forgotten about them.

However, the sightings came back with a vengeance, particularly after the Soviet Union's launch of the second Sputnik in November and as American satellites sat on the launching pad. Newsrooms were besieged with reports. *Winners and Sinners*, the internal bulletin of the *New York Times* newsroom for November 21, 1957, warned: "Flying saucers streamed in the wake of the Sputniks and reports of them are sure to recur."

The "Spooknik" phenomenon added to the mix. Spooknik was the name

A gigantic balloon five minutes before its ascent on a record-breaking flight into the stratosphere 13.71 miles above sea level. (NASA)

given to a ghost signal—a radio doppelgänger—which seemed to come down from the empty sky when Sputnik was on the other side of the world. It was quickly explained as a "hot spot" in the ionosphere reflecting the satellite, but the idea of a ghost satellite fueled the notion that something odd was going on overhead.

Also adding to the mix was the sudden rise in the number of homemade rockets in the sky. Countless groups of model-building kids, weaned on Buck Rogers, Tom Corbett, and *Popular Science,* were firing off such rockets throughout the United States.* This became such a problem in Indiana that, on December 10, they were actually banned. A day later, the state of Connecticut held a parley of insurance representatives, police, and fire officials to deal with the problem.

Never before had so many people looked skyward with a purpose or for so long. Of course, they saw more on some nights than on others. The

*Some of those backyard rocketeers would later become part of the American space program. In his Capraesque memoir *Rocket Boys,* which became the movie *October Sky,* Homer H. Hickam Jr. recalled using scrap metal from West Virginia coal-mining equipment to create spectacular post-Sputnik rocket launches with his pals. Known as the Big Creek Missile Agency (BCMA), the boys went on to win science awards and scholarships. In 1997, just before he retired from a career at NASA training astronauts, among other duties, Hickam got an astronaut friend to carry a piece of a rocket and a Science Fair medal into space. He declared at the end of his book that the BCMA was finally in space.

biggest night would come the next year. Shortly after midnight on April 14, 1958, UFO sightings were reported by reliable witnesses along the east coast of the United States, especially in Connecticut and on Long Island.* They reported a brilliant, bluish-white object moving high across the sky at an incredible speed. According to the reports, it suddenly turned red, and several smaller objects detached themselves from the main object and fell into formation behind it.

Minutes later, observers on more than fifteen ships in the Caribbean reported seeing one or more brilliant objects of different colors and configurations, all moving in the same direction in the sky. Some saw the fleet, others saw a single object with an enormous tail, and a few saw both—such as this witness from one of the ships: "The main body appeared to have a blue-white head, then a short dark space before the glowing orange-yellow tail. Twenty-seven separate particles were actually counted as they appeared in the main plume. Each followed the main body and each developed its own glowing tail on leaving it." Witnesses on the ground from the Virgin Islands to British Guiana saw the fleet of objects, which to many looked like an alien invasion led by a massive mother ship and her satellite craft.

What these hundreds of people were actually seeing was the flaming death of Sputnik II, the vehicle carrying the corpse of Laika. It reentered Earth's atmosphere on April 14, 1958, after 162 days in orbit.†

The true believers refused to see the Sputniks as "explainable sightings" but rather as actual UFO sightings. The periodical *Flying Saucer Review* alleged that since it was "highly unlikely" that the Sputniks could actually be seen with the naked eye, it was more likely that what people were actually seeing were UFOs themselves. It was even claimed there were so many sightings because UFOs were attracted by the Sputnik launches and had come near Earth for a look.

Over time, the 1957 Sputniks emerged as seminal events in the in-

*Two of these witnesses were my in-laws, who saw a UFO from their patio in Long Island, New York.

†The true believers hate to admit it, but since the Sputnik II reentry a lot of space debris gets labeled as UFOs. As recently as September 1, 1999, the remains of the engine from a Russian SL-12 booster—one of the 8,765 man-made objects then in Earth orbit—came back and burned up in the atmosphere over the Pacific Northwest. "Something's going on out there—something big," reported Art Bell on his overnight radio talk show. According to the Portland *Oregonian* of September 24, 1999, Bell, the National UFO Reporting Center, and others insisted it was "big news and evidence of a government cover-up."

creased popularity of UFOs. The original Sputnik achieved a unique place in the lore of UFOs and abductions by extraterrestrial beings. In mid-October 1957, within days of its launch, came the first report of an alien abduction, although it did not appear in UFO literature for another five years. The event occurred in Brazil, where, according to Peter Hough and Jenny Randles, authors of *The Complete Book of UFOs*, one Antonio Villas Boas was relieved of his seed through forced copulation with a "space woman" who had a pointy chin.

Sputnik provoked a whole new phase of the UFO phenomenon, which involved harvesting sperm and other genetic samples from humans for what was reputedly the creation of hybrid space babies. A large cohort came to believe that the Sputniks served as a mating call to abductors and that no lonely, isolated human has been safe from interplanetary molestation ever since.

The Complete Book of UFOs and other sources also recount that on November 18, 1957, a twenty-seven-year-old mother named Cynthia Appleton from Birmingham, England, heard a high-pitched whistling noise, smelled something like ozone, and saw a rose pink hue spread throughout her suburban home. Out of the hue materialized a tall humanoid creature with elongated eyes, pale skin, and long blond hair. He wore a silver one-piece suit with a covered helmet. Cynthia had a telepathic chat with the alien, who told her that he was from a planet called Gharnasvarn, which wanted to make peaceful contact but hesitated because of the Earth's atomic weapons. He made eight more visits—sometimes driving to the house and once wearing a homburg hat—and finally told her that she would have a cosmic child whose father would be her husband but who would actually belong to the race on Gharnasvarn. The luckless husband, a welder named Ron, said that if the child was really the product of some kind of alien immaculate conception, the folks there should pay for them to raise the child. Cynthia sent a telepathic message and was told the Gharnasvarnians do not use money.

CHAPTER EIGHT

American Birds

> The successful orbiting of Explorer 1 is one of the landmarks in the technical and scientific history of the human race. Its instrumentation revealed the existence of radiation belts around the Earth and opened a massive new field of scientific exploration in space. It inspired an entire generation of young men and women in the United States to higher achievement and propelled the Western World into the space age.
>
> —Astrophysicist James A. Van Allen, speaking on March 31, 1970, when Explorer I reentered the Earth's atmosphere

State University of Iowa astrophysicist James Van Allen, who was chairman of the Working Group on Internal Instrumentation of the Technical Panel Earth on Satellite Program, was in charge of IGY's scientific payload that was to be sent into space by the United States. When Vanguard became America's official choice for the IGY, Van Allen had joined Vanguard's team.

Ernst Stuhlinger, of the Army Ballistic Missile Research Center in Huntsville, Alabama, was sent to Iowa to enlist Van Allen in the effort to get the Jupiter C selected as the IGY vehicle, even though work was progressing on the Vanguard. Van Allen studied the information and concluded that Stuhlinger was right. He sent telegrams to all the other members of the technical panel, urging them to drop out of the Vanguard program and join the Jupiter C–powered Orbiter mission. Van Allen, who was turned down by the panel, decided to design his package so that it would fit into either the Vanguard or Orbiter satellite—just in case.

In Huntsville, Alabama, the Rocket Team still faced the dilemma of locating a satellite and an experimental payload for its Orbiter mission. They had to work quickly and quietly because they were preparing for a launch without permission, let alone financial authorization.

Locating the satellite was easy. William H. Pickering, director of the California Institute of Technology's Jet Propulsion Laboratory in Pasadena, had been in a state of semi-readiness for some months, waiting for the green light to complete the upper stages of the Jupiter C rocket, including the final, fourth stage: the satellite that would actually go into orbit around the Earth.

Work on the Orbiter proposal continued on both coasts even though Washington officials had canceled it and given the order to stop all work. Those involved in the deception called it "bootlegging," Albert R. Hibbs, a space science pioneer, later admitted: "At JPL, we had for several months been building the satellite for Orbiter. We had a difficult time continuing, and bootlegged the whole job under the title of a re-entry vehicle program. When finished we locked up the satellite in a cabinet up in [the] solid-motor test area so it wouldn't be found."

The satellite itself was no longer called Orbiter but rather Deal. When Orbiter was turned down, John Small, a section chief at JPL, made a pronouncement that is familiar to poker players: "The winners laugh and joke and the losers yell 'deal.'" The name stuck. While waiting to hear from Van Allen as to whether he would be willing to make the switch from Vanguard, Pickering and Hibbs went to Washington to get permission to continue work on Deal and to install Van Allen's experiment package inside.

For a number of days neither the Navy nor the Coast Guard was able to find Van Allen. He was at sea aboard the Navy's USS *Glacier* en route to Antarctica, where he was going to study cosmic rays as part of the IGY. On October 29, a desperate Pickering sent the following telegram:

> DR. JAMES A. VAN ALLEN, USS GLACIER, SOUTH PACIFIC. WOULD YOU APPROVE A TRANSFER OF YOUR EXPERIMENT TO US WITH TWO COPIES IN SPRING, PLEASE ADVISE IMMEDIATELY. W. H. PICKERING.

Although Van Allen had some trouble understanding the request, he replied by cable, giving Pickering a blanket OK to do whatever he thought was best. Within twenty-four hours one of Van Allen's assistants, a graduate student named George Ludwig, was on his way to Pasadena with the payload—a micrometeorite detector in the form of a tape recorder and a cosmic ray experiment.

Official permission to make the transfer was given in Washington on November 22. Pickering pointed out that the Army satellite, which had been designed to spin for stability, would not allow for the flight of the tiny tape recorder, which was part of the Vanguard specifications. The satellite experiment package was readied for the stovepipe configurations of the Army's satellite instead of the Navy's ball-shaped Vanguard.

The Launching of Juno

The actual launch vehicle that would put the first American object in space was sent by plane from Huntsville to Cape Canaveral (the same location where the next Vanguard Test Vehicle was being prepared for launch). On December 20, 1957, the ABMA Missile Firing Laboratory, headed by Peenemünde alumnus Kurt H. Debus, began preflight tests for Jupiter C and prepared the main stage for firing. This was accomplished under heavy guard in a hangar at Patrick Air Force Base. Wernher von Braun, director of development at the Army Ballistic Missile Agency in Huntsville, Alabama, and his boss, Major General John B. Medaris, who had promised to put an Army satellite in orbit within ninety days, were moving according to schedule.

The Jupiter C, which had begun its life as a Redstone, was renamed the Juno because Juno was the sister and wife of the Roman god Jupiter; the Juno rocket would be a satellite-bearing sister of the Jupiter C. The immense modified unit that arrived at the cape was a Juno carrying the large letters *UE*, which meant that it was Redstone 29. In an attempt to keep secret how many Redstone missiles were actually being produced, the Army came up with a system that owed more to the Captain Midnight secret decoder ring than to modern cryptology. Each letter in the name Huntsville represents a digit.

H = 1 U = 2 N = 3 T = 4 S = 5 V = 6 I = 7 L = 8 E = 9 X = 0*

*Hence the first Jupiter C launched using Redstone 27 as its first stage on September 20, 1956, carried the number-letters *UI*. The simplicity of the code notwithstanding, it broke down with Redstone 49. Labeled *HE*—it should have been labeled *TE*—Redstone 49 was launched on August 24, 1958.

On December 20, the Juno rocket was transported to Cape Canaveral's Hangar D to undergo additional tests and calibrations. It was erected under the cover of darkness on January 15 without searchlights at Pad A of Launch Complex 26. Scaffolding was put up, and the spin-stabilized upper sections were put into place. Most of the vehicle was shrouded in heavy canvas. On December 24, the final upper stages were added, including the satellite.

The ABMA's opportunity arrived on January 26 when the backup to the ill-fated TV-3 vehicle, the Vanguard TV-3BU (i.e., Test Vehicle-3 Back Up), had to be delayed pending a second-stage engine replacement. Twice it had come to within seconds of a firing—once to within twenty-two seconds, once within fourteen—and then the problem-plagued Vanguard simply lost its place in line.*

The Huntsville team then went to the front of the line for five days. Beginning on January 28, at about the time they were finally ready to launch, weather balloons sent aloft detected a severe shift in the weather, so the first scheduled launch was scrubbed. The seventy-foot-long space vehicle was susceptible to wind shear, and it was feared that it could break up or tumble out of control. A check the following morning showed worsening conditions, and a second countdown was scrubbed later that day as high-altitude winds reached a speed of 217 miles per hour.

Things looked bad for the Army on the night of January 30, after the second cancellation. The second Vanguard firing had been rescheduled for February 3, and missiles couldn't be launched at the cape at intervals shorter than three days. If the Army did not launch its vehicle by the next night, it would have to move over for Vanguard and then for a series of urgent Air Force Missile tests.

During all of this time, the weather on the ground seemed ideal—sunny with a light breeze—giving rise to the suspicion among the assembled press corps that something was wrong with the hardware. Fortunately, the public knew none of this because the Army, mindful of the Vanguard experience with advance publicity, guarded the launch date until the last minute. President Eisenhower had been adamant about not repeating the Vanguard publicity

*One final bad omen for Vanguard appeared in the night sky. During that twenty-two-second hold on one of the coldest January nights on record in Florida, a newsman suddenly shouted to those assembled for the launch to look up. It was Sputnik II passing overhead, like "a cold and glowing celestial ghost," as one of those present put it. Wags at the cape suggested that there would be a Sputnik traffic jam before America got into space.

debacle. This time the president himself would make the first announcement of the launch *only* if and when it was in orbit.

The press corps at the cape, now about eighty-five strong, realized the competition for lurid headlines and sensational coverage of the Vanguard explosion had made matters worse. They had been willing to strike a bargain with Air Force major general Donald N. Yates, the commanding officer at Cape Canaveral. According to a report in *Newsweek*, Yates agreed to give advance information on scheduled launchings if reporters would avoid dramatizing inter-service rivalry and write no advance stories on upcoming launches. According to the agreement that was struck, the story could not be told until there was "fire in the tail" of the booster. This amounted to self-censorship, and it worked; the closest thing to a leak was a note in the local *Cocoa Beach Tribune* on January 29. The news corps at the cape was betting that the Army would attempt a launch "this week."

On January 31, the jet stream shifted north; conditions were better but still marginal. A young Air Force meteorologist, Lieutenant John L. Meisenheimer, determined that the wind shear would subside long enough to launch, but it was likely to return within a few hours. Concerned about the effects of further delay on the efficiency and morale of his firing group and the loss of the launchpad to the Vanguard, Major General Medaris bet that Meisenheimer was right and decided to give the final order to launch.

A few last-minute hitches were overcome, and finally the destruct package was armed; it would destroy the missile in flight if it went off course. As final preparations were made, odds were posted ranging from one in ten to the 90 percent certainty predicted by Medaris. Major General Yates put the odds at one in three. Reporters called home to reserve room on the front pages of their papers' Saturday morning editions.

At 10:15 P.M. gigantic floodlights were turned on, bathing the snow-white Juno in a cone of light. Red warning lights blinked, Klaxon horns sounded, and white fumes of liquid oxygen billowed out like steam from an enormous kettle. Atop the missile were the final stages of the vehicle housed in a large spinning tub that stabilized the satellite itself. (The satellite was still called Deal. The poker player's metaphor was validated: The cards had been dealt and were now on the table.)

At exactly 10:48 P.M. eastern time on January 31, 1958, the firing ring was pulled. Eleven seconds later, the last connection hose was sliced free, and ignition occurred one second later. A huge firelick burst from the base of the missile and expanded into a gigantic orange ball of flame. There was an earthshaking roar, and at 10:48:16 Juno was airborne. Reporters lost all sense of decorum: They waved their arms, shouted, pounded each other on the back.

Ninety seconds later, Juno went through the jet stream, suffering no ill effects from the wind.

Using slide rules, precomputed charts, and a crude mechanical device called an Apex Predictor, Command at the blockhouse sent the order for the first stage to separate and the second stage to fire. The second stage took the satellite up to 9,000 miles per hour; the third stage ignited, pushing the speed to 15,000 mph. Then the final, or "kick," stage put the satellite into an orbital speed of 18,000 mph. Seven minutes after that first lick of flame, the needle-nosed satellite entered its orbit in space. At this point the spinning of the satellite caused whiplike rods tipped with weighted metal balls to fan out. These were the antennae that would allow the satellite to talk to ground stations and announce its position.

Pickering, Van Allen, and von Braun were not at the cape but in Washington monitoring the event at the Pentagon. Much to von Braun's displeasure, they had been sent north for one reason and one reason only: to meet the press. "We had been told in so many polite words that the public information aspects . . . were far more important than the firing itself—so we had to sweat it out in the Communications Room of the Pentagon," von Braun told an audience on March 18, 1958. He added: "Now I will let you in on a little secret. I was told that if everything worked successfully, we would go over to the National Science Academy to meet the press, newsreel and television people. That's why I put on a dark suit. But, just in case things didn't come off so well, I had a pair of dark sunglasses with me and was determined to sneak away to a still darker movie theater."

Seven minutes after the launch, about the time the satellite and the attached fourth stage went into orbit, Army secretary Wilbur Brucker, who had come to the Pentagon from a dinner party to join von Braun and company to monitor the satellite, sent a teletype to Medaris at the cape:

WHAT DO WE DO NOW?

Medaris typed back:

HAVE A CIGARETTE AND SOME COFFEE AND SWEAT IT OUT WITH US.

Over the next hour, several reports came in to the Pentagon that U.S. military stations had picked up a "peeeeeeep," the steady seven-note harmonic signature of this satellite. It was transmitting loud and clear.

"The time dragged," according to the notes of Major General John H. Hinrichs. "More cigarettes, more coffee, more doughnuts." Then, "after what seemed to be a hundred hours instead of minutes, Bill Pickering charged into the room saying 'I can't stand this! I'm going to get Pasadena.' So he got on the phone with JPL ([Explorer project manager Jack E.] Froehlich, I think). But there was no information there yet, so there was some long-distance chatter on what was going on at either end of the line. At about 106 minutes—the first estimate for [the satellite] to be over San Diego—Bill started prodding for the answer to 'Have you got it yet?' This went on forever it seemed. At one point a report came into von Braun (on another phone) from NAS, where the IGY crew were waiting, that Earthquake Canyon (monitor station) had it.

"Pickering queried JPL—'No such thing! None of us have it!' "

Pickering then asked, "What do we do now—go out and cut our throats?"

Von Braun responded, "False report."

By this point, Hinrichs recalled, "Bill was squirming in his chair, burning his watch up with his intense gaze, and then suddenly sat up. 'Earthquake has it?' (Pause) 'You have it?' (Pause) 'Temple City has it! It *is* in orbit! Phew!' " The word was passed, everyone congratulated everyone else—pictures were shot. Jubilation!

The Orbit Is Confirmed

At 12:51 A.M. on February 1, about 100 minutes after liftoff, a successful orbit was confirmed by the Jet Propulsion Laboratory in Pasadena, and loudspeakers at Cape Canaveral and the Pentagon telecommunications room gave the final signal that the Army Tracking Station in Earthquake Valley, California, had "the bird."

Deal had come around the world and would soon pass over the East Coast of the United States.

A few minutes later, Secretary Brucker called Medaris at the cape to tell him that he and Maxwell Taylor, Army chief of staff, had selected the name Explorer for the Army satellite.*

President Eisenhower, on a golf vacation in Augusta, Georgia, was

*All sorts of names were suggested: Medaris had wanted to call it Highball, Army secretary Brucker favored Top Kick, and a contingent from JPL wanted to hold on to the name Deal.

William H. Pickering of the Jet Propulsion Laboratory, James A. Van Allen of the State University of Iowa, and Wernher von Braun of the U.S. Army's Redstone Arsenal triumphantly holding aloft a model of Explorer I at a news conference early on the morning of February 1, 1958. (NASA)

called by Press Secretary Hagerty at 12:44 A.M. in the midst of a late-night bridge game. When told that the satellite was in orbit, Eisenhower said, "That's wonderful. I sure feel a lot better now." Then he added, "Let's not make too big a hullabaloo over this."

A minute later, Hagerty put out a prepared statement under Eisenhower's name: "The United States has successfully placed a scientific Earth satellite in orbit about the Earth. . . . This launching is part of our country's participation in the International Geophysical Year. All information received from this satellite promptly will be made available to the scientific community of the world." No mention was made of the Army in his announcement, but anyone who had been following the story knew who was responsible.

At 2:00 A.M. Pickering, von Braun, and Van Allen attended a press conference and photo opportunity at the National Academy of Sciences on Constitution Avenue. They celebrated the event by hoisting a model of the cylindrical eighty-inch-long Explorer satellite above their heads. It was a grand-looking silver-colored object, marked with dark brown stripes running down the sides.* Its total weight was 30.66 pounds, of which 18.35 were in-

*The actual satellite in orbit had been sandblasted until it took on a natural black appearance, and the stripes were white. The model of Explorer I on exhibit at the Kennedy Space Flight Center Visitors' Center replicates the black model that went into orbit.

struments. The stripes, they explained, were layers of zirconium oxide applied to minimize reaction to temperature changes. The mood inside the room was delirious.

The triumphant image of the three appeared in virtually every newspaper in the United States and many throughout the world. In Huntsville, news of the successful mission was greeted late that night with "the wail of sirens, blaring horns and the fiery trails of store-bought rockets," according to the *Huntsville Times*'s "Satellite Extra" of February 1, 1958. Ten thousand people celebrated in the streets. Mayor R. B. Search proclaimed, "It is the greatest day in the history of Huntsville."

There also was momentary anger. Former secretary of defense Charles E. Wilson, the man who slowed the Army missile program and once came close to closing down the Redstone Arsenal, was burned in effigy. When told this news while on vacation in Miami Beach, Wilson responded: "I don't know why they're mad at me. I'm the one who put them in the Jupiter business. They must have me mixed up with the Ku Klux Klan."

As with Sputnik I, the news of Explorer was a weekend story, giving Americans time to savor the event. The Saturday morning headlines were bold and oversized: "First U.S. Moon Circling Globe" (*San Francisco Chronicle*), "U.S. Moon Hailed by World, Sends Signals Perfectly" (*Chicago Daily News*), and "Moon over the Cape" (*Orlando Sentinel*).

America was now in the race, and everyone wanted to share in the glory. The *Los Angeles Examiner*, for instance, carried the headline "Caltech Satellite Circling Earth," while the *Los Angeles Times* proclaimed in one of its headlines: "Launching Is Tribute to Southland Science."

Some pressure was off the White House and the Republicans. A group of proud GOP members of Congress released prerecorded films explaining the American space achievement for television viewers in their districts. The only problem with these films was that they showed the spherical Vanguard rather than the bullet-shaped Explorer.

Putting a nice but also inaccurate spin on things was Vice President Nixon, who linked Explorer with "President Eisenhower's proposal for the development of space exploration in the cause of peace rather than the wastage of war." Juno was, after all, an intermediate-range ballistic missile sired by the V-2.

The Army, whose officers and civilian scientists had gone out on a limb, was given public vindication. A *Newsweek* cover story, titled "Our Moon: How the Army Came Through," noted that "probably never before in peacetime had the Army attained such wide popularity."

The Army in general and ABMA in particular had now legitimately achieved leadership in space and were not afraid to say it. "It is a highly balanced team," said Major General Medaris immediately after the launch, "which will make other notable contributions in the interest of science and the national defense." According to Medaris, the team was the Army, American science, and American industry. Von Braun said, "This is the beginning in the long-range program to conquer outer space."

Organizing for Space

The next few weeks saw a flurry of action in Washington, D.C. It began on February 4, 1958, when President Eisenhower directed his new science adviser, James R. Killian Jr., to head a committee to study and make recommendations on the governmental organization of the nation's space program. Explorer's success encouraged supporters of a crash effort to recoup lost U.S. prestige by launching an automated probe to the Moon. The idea, first discussed the morning after Sputnik, came up on February 4 at the Legislative Leadership Meeting at the White House—an opportunity for Republican congressional leaders and the Eisenhower administration to compare notes. Ike was cool to the idea. A written synopsis of the meeting said the president was "firmly of the opinion that the rule of reason had to be applied to these Space projects—that we couldn't pour unlimited funds into these costly projects where there was nothing of early value to the Nation's security. . . . in the present situation, the President mused, he would rather have a good Redstone than be able to hit the moon, for we didn't have any enemies on the moon!"

When Senator William Knowland pointed out the prestige value of being first to hit the Moon, Eisenhower relented somewhat, saying that if a rocket now available could do the job, work should go ahead. The president stressed, however, that he "didn't want to rush into an all-out effort on each of these possible glamor performances without a full appreciation of their great cost."

Meanwhile, a second but less spectacular Vanguard failure occurred the next day, February 5, 1958. This time, the malfunctioning rocket (TV-3 BU), which had been forced off the launchpad for Explorer, broke up a minute into flight at 20,000 feet.

The next day, the Senate formed the ad hoc Special Committee on Space and Astronautics, chaired by Lyndon Johnson. Seeds that would lead to the creation of a civilian American space agency were germinating quickly. Two weeks earlier, the Senate Preparedness Investigating Committee under Senator Lyndon B. Johnson had summarized its findings in seventeen specific recommendations, including the immediate establishment of an independent space agency.

On Sunday, February 9, 1958, Major General Medaris appeared on *Meet the Press*, which was then broadcast on both radio and television. He was asked by John Finney of the *New York Times* how much sooner he could have put an Explorer satellite in orbit if he had been given Defense Department approval over Vanguard. "I think there is little question but we could have put it up about a year sooner than we did," he replied.

Medaris was acting as a free agent, again second-guessing Washington and straying into uncharted waters. When asked if he were ordered to hit the Moon with a missile, how long it would take, he said "months." Medaris would not say when he would put up his next satellite other than to predict, "A matter of weeks. Sometime between tomorrow and the first of April, I should say."*

On March 5, a second Explorer launched but never made it into orbit because of a fourth-stage rocket malfunction which caused it to crash back to Earth. The experiment-laden satellite it carried was lost. For the first time in months, the voices of Medaris and von Braun were muted, and they shied away from the press coverage that they had worked so hard to attract. The Army had put one in the loss column.

Also on March 5, a presidential advisory committee recommended to Eisenhower that "leadership of the civil space effort be lodged in a strengthened and redesignated National Advisory Committee for Astronautics (NACA)." This jibed with the wishes of Lyndon Johnson and his subcommittee and set the course in the direction of a civilian space agency.

*On the lecture trail, Medaris advanced his own apocalyptic view of the world. On March 24, 1958, while addressing an aerospace group in Pittsburgh on the topic of "Army Missile Progress," he turned to the subject of space and mixed in a little brimstone: "The overriding importance of space exploration takes on proper significance only if we appreciate that it will make possible a new understanding of man's relationship with the infinity of Divine Creation. If we fail to read the signs right, we may find progress preempted by the agents of the anti-Christ with awful consequences." He tossed off such aphorisms as "government must conform to God's will."

Vanguard Is Exonerated

After four scrubbed launch attempts, Vanguard finally rose into a sunny sky on March 17, 1958, at 7:15:41 A.M. The satellite went into Earth orbit, along with the burned-out third-stage rocket casing. A total weight of about fifty-five pounds was placed into orbit, including the four-pound payload. In his cubicle of an office in Washington, John P. Hagen got the good news and put in a phone call to Alan Waterman, director of the National Science Foundation, telling him he should inform the president that Vanguard was in orbit.*

Eisenhower was delighted and was quick to congratulate all concerned. The president sent his old friend Admiral Arleigh Burke, chief of naval operations, a bottle of Burke's favorite scotch, Chivas Regal, along with a note. Burke responded with a handwritten letter, which started out coherently but ended, "Mush quitnow and fine anodder bodel odish delicious boos." Eisenhower thought the letter was hilarious, carried it in his pocket for weeks, and showed it to anyone he could corner.

The White House clearly saw the Navy's Vanguard as vindication of its own deliberate, open, scientific approach to space. Vanguard I was the second U.S. satellite and the world's fourth. It circled the globe every 107.9 minutes in an elliptical orbit, with an apogee of 2,466 miles and perigee of 404 miles. Despite its small size, it could be traced optically.

The immediate Vanguard success was met with a pat on the back and the kind of trivial response reserved for also-rans.† Nikita Khrushchev compared it to Sputniks I and II and scoffingly called it a grapefruit. That label became so popular that it made its way into the *World Book Encyclopedia*'s yearbook for 1958, which noted: "Rocket experts called it 'the grapefruit.' "

*The launch was flawless. The only mishap of the night occurred after Vanguard was up and the launch team waited in the teletype room for confirmation that the satellite had made its first orbit. The room was packed so solidly that Stehling compared it to the ship's cabin in the Marx brothers' film *A Night at the Opera*. The overworked coffee machine, which had been running at full tilt for many hours, literally blew its top, spewing coffee all over the room. Nobody seemed to care, because moments later the orbit was confirmed.

†One congressman wanted it officially renamed the St. Patrick's Day satellite because it was launched on March 17. When word leaked that a St. Christopher's medal had been precisely wired to its gyro package, a minor but very real "separation of church and state" commotion was touched off. This Roman Catholic intrusion was denounced from a few Protestant pulpits, and it caused some hand-wringing in Congress.

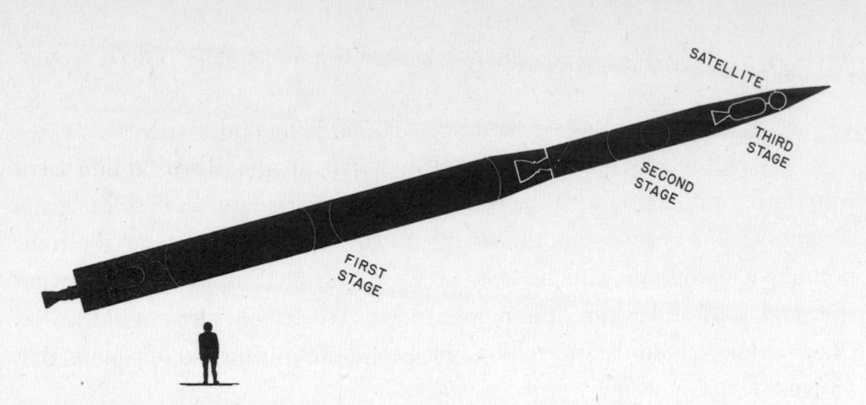

Diagram of the three-stage-rocket Vanguard launching vehicle. (U.S. Navy)

Comparing Vanguard's launch vehicle to the vehicle that had sent Explorer aloft was akin to comparing a surgeon's scalpel to a woodsman's ax. The Vanguard rocket was designed as a slim and efficient vehicle of scientific exploration, not an instrument of mass destruction. Standing seventy-five feet tall, it had a diameter of only forty-five inches, giving it the proportion of an enormous pencil rather than a missile. Its Vanguard payload was tiny, lean, and long-lived; it was meant to stay in space for a number of years, as compared with Explorer's lifespan of just over eight weeks.

The Vanguard satellite was extremely successful in many ways. It pioneered the use of solid-state devices, printed circuits, and the general principle of miniaturization. The idea was to create as much scientific power as possible within the smallest space. This was a momentous achievement at a time when a computer was the size of a room.

The satellite was the first to use solar cells to create power. There were two transmitters aboard the satellite: One was powered by conventional batteries, and the other used solar cells. The solar-powered transmitter was simplicity itself—thin layers of silicon that when exposed to sunlight absorbed the Sun's energy and transmitted a small electric current, a mere 0.06 watts, but sufficient to activate a radio beam. The satellite was also placed into almost perfect orbit far above the denser part of the atmosphere, where there is little atmospheric drag.

Vanguard eventually provided data showing that the concept of Earth as an almost perfect sphere with a slight flattening at either pole is wrong. Vanguard gave proof of what had only been theorized before: that the Earth is slightly pear shaped—a discovery that has been useful in correcting errors in geographic maps. The pear-shaped distortion causes the sea level to be about 100 feet higher at the North Pole than the South Pole. This discovery was made as trackers using the solar-powered signal were able to measure the satellite's distance from the Earth in exceedingly fine detail.

Still another discovery made by Vanguard I was that the orbit of a satellite is influenced by solar wind, the pressure of sunlight, the Earth's magnetic field, the atmosphere above 400 miles, and the gravitational fields of the Moon and Sun.*

Vanguard's original 200-year life expectancy was much too conservative. Vanguard today remains the oldest man-made object in space, expected to orbit Earth for at least 1,000 more years and perhaps as many as 2,000.

"This launching occurred two years, six months and eight days after the initiation of the program from scratch," its director, John P. Hagen, wrote later. "Vanguard started with virtually nothing in 1955, completed vehicle design in March 1956, and had a fully successful flight two years later. One can challenge any other new rocket program in the United States to demonstrate a completely successful launching within such a short time."

On March 26, less than a month after Vanguard I's launch, the 30.66-pound Explorer III went aloft with an 18.83-pound payload, and America seemed to be on a roll with three satellites in orbit.

The United States was also experiencing an international public relations victory on terra firma. In April 1958 Harvey Lavan Cliburn Jr., a relatively unknown twenty-three-year-old Texan, won the first Tchaikovsky Competition for pianists. Americans perceived this as beating the Russians at their own game—music—and on their own turf. Van Cliburn played not only the required Tchaikovsky *First Concerto* but works by two other Russians (Rachmaninoff's *Third Piano Concerto* and a rondo by Dmitri Kabalensky). The Russians fell for the tall blond, and he was showered with gifts and adoration. He visited with Premier Khrushchev.

Van Cliburn came home to a hero's welcome unlike any the country had

*Yet the highly publicized and telegenic Vanguard failure on December 6, 1957, created a lasting negative impression of a spectacular and public propaganda catastrophe. The film of that first sensational explosion has become part of virtually every television documentary on the space race or the Cold War. When Hagen passed away in 1990, the obituary writers remembered him more for his failure than for his success. The *New York Times* called him "a leading figure in the dramatic and disappointing early days of the American space program." After recounting his failures, they mentioned the three-and-a-half-pound satellite that is the oldest man-made object in space, but there was no discussion of its many successes. The errors are not of fact but of omission. For years, critics dwelled on its price tag. "An expensive fiasco for the most part" is how Isaac Asimov later dismissed the Vanguard program.

given in peacetime since Charles Lindbergh's in 1927. He got a ticker-tape parade down New York's Fifth Avenue; was received at the White House by President Eisenhower; and bobby-soxers reacted to him with an ardor usually reserved for pop-music idols. Writing about "Vandemonium" in 1994 in the *San Diego Union-Tribune*, critic Welton Jones noted, "For that time and place he was bulletproof, a full set of Teflon-coated attitudes and achievements politically correct decades before the concept was labeled. After all, it was Van who paid back the Russians for the insult of Sputnik."

The Tortoise Becomes the Hare

The Soviet Union planned to have its heavily instrumented third Sputnik in orbit for May 1, or May Day, the celebration of international socialist solidarity. When it was launched on April 27, there was a rocket engine failure at about twelve kilometers, and the satellite was lost.*

The Russian failure was not known in the West for many months. But it had symbolic importance for Americans because on the same May Day James Van Allen had announced that radiation detectors aboard Explorer I and Explorer III had led to the discovery of powerful radiation belts surrounding Earth.

The detection of the Van Allen belts was the first major space discovery and the most important finding of the IGY. A *Time* magazine cover story called him America's foremost space scientist and a "key figure in the cold war's competition for prestige."

Supporters of Eisenhower's methodical approach to space exploration capitalized on the Van Allen find, pointing out that the Soviet Union's two heavy Sputniks had accomplished no equivalent scientific feat. Never mind that, according to Major General Medaris, Van Allen's discovery was sheer accident. "[Explorer] was supposed to go into a 125-mile circular orbit," Medaris explained on the anniversary of the tenth year in orbit. "We didn't have the attitude controls of today, and the satellite got tipped up a bit, going into an elliptical orbit ranging from 224 to 1,573 miles. The upper part of the orbit penetrated the belt."

*Earlier accounts suggested that the rocket and satellite fell separately back to Earth and that after a frantic hunt the satellite was found and brought back largely intact to the launch site in an armored vehicle, but this was refuted by Sergei Khrushchev in an essay that appears in the book *Reconsidering Sputnik* published in the year 2000.

The impression left for posterity was that all of the delicate engineering that went into the Van Allen experiment had been done overnight, as if a rabbit had been pulled out of a hat. The press gave no hint that the Vanguard team had spent months creating reliable temperature controls, testing instruments, and putting it all together in an ingeniously compact and efficient package before the experiment was turned over to Explorer. "So far from realizing that Vanguard had contributed anything to the later success of Explorer, much of the American public thereafter assumed that the Army, notably Wernher von Braun with a major assist from JPL, had prepared the Explorer payload from start to finish and in less than eleven weeks," concluded Constance McLaughlin Green and Milton Lomask in their book *Vanguard: A History*.

By the same token, nobody at the time publicly raised the question of how JPL had created a satellite in a matter of weeks, along with a photogenic copy to carry to a press conference. To the present day, there are otherwise authoritative sources that report the fully instrumented satellite being created from scratch in short order.

The next four attempts to orbit Vanguard satellites failed, but the Vanguard II satellite was successfully launched on February 17, 1959. Its mission was to examine weather conditions. It developed an unfortunate tumble rate in orbit after being nudged by its own rocket casing, but it did establish the feasibility of the principles on which weather satellites could be designed. It is expected to remain in orbit for several hundred years.

After two more Vanguard failures, the very last launch of a Vanguard scored a success. The Vanguard III satellite was successfully launched on September 18, 1959, with a fifty-six-pound payload. It mapped the Earth's magnetic field and returned data on the lower edge of the Van Allen Radiation Region. Vanguard III went aloft on the last of the seven launch vehicles built under Navy aegis for the IGY, and as NASA later decided not to commission any more, Project Vanguard came to an official end shortly after this flight.

Part of the legacy of the Vanguard rocket is its contribution to the genesis of the U.S. space program. Its upper stages formed the basis for upper-stage configurations employed on Atlas-Able, Thor-Able, and Scout rockets. Vanguard technology also helped the NASA Apollo program, with its modified upper stage forming the basis of the Atlas-Antares second stage. Atlas-Antares was used to test various proposed Apollo reentry vehicle designs.

Unlike the Vanguard, the Explorer program lives on to the present day. Explorer was inherited by NASA from the Army and enjoyed a long life, extending through a fifty-fifth incarnation, which was sent up in 1975. In honor of the original Explorer I experiment, all cosmic ray experiments carry the Explorer name. In 1997, the Advanced Composition Explorer (ACE) was sent out one million miles to study low-energy solar radiation and galactic radiation.

The Juno family tree is just as impressive. There would be three more attempts to launch satellites aboard Juno I rockets after Explorer III. The first was successful, carrying Explorer IV into orbit, but the last two launches failed. On August 24, in an attempt to orbit another Explorer, the booster bumped into the upper stages of the rocket, sending it out of control. On October 23 the upper stages of a Juno broke free of the booster, failing to orbit the Beacon 1 satellite.

With a 3–3 record the Juno was retired, yielding to more powerful rockets beginning with the Juno II. The next models—Junos III and IV—never left the drawing boards, but the V rocket was transferred to NASA and renamed Saturn. The descendants of the Junos of 1958, sired by the V-2, Redstone, and Jupiter, would send men to the Moon.

Reduced to basics, the Army's efforts developed into the manned spaceflight program, while Vanguard laid the groundwork for space science and the unmanned space program. The Vanguard satellite was a great starting point for an unmanned satellite program. It was lightweight and used highly miniaturized electronics yet was still large enough to be tracked optically. Within it was a Minitrack transmitter—the Minitrack system and network would be used by the United States for all Earth satellite tracking for many years to come. After October 1958, Vanguard became part of NASA, and a substantial amount of the $110 million that had been allotted to the program was turned over to the new agency.

U.S. Reaction to the Immense Sputnik III

Korolev's team had not stopped work since Sputnik II. On May 15, 1958, Korolev finally launched the conical, 1,330-kilogram (2,926-lb.) Sputnik III satellite. This gargantuan third Sputnik caused less of a public furor but more of one within the inner circles of government and science because of its size and function. Indeed, the combined weight of

Model of Sputnik III, which was launched on September 18, 1958, earning a new level of respect from the United States for its far greater size and heavy instrumentation. (Author's collection)

the first three U.S. satellites was only a tiny fraction of the weight of Sputnik III. In his book *Race to Oblivion: A Participant's View of the Arms Race,* Herbert F. York notes that "these huge differences in weight served to worsen considerably the general consternation that pervaded all levels of American society and government. It was widely felt that the Russians must be onto some secret that had evaded us."*

What the West did not know until the 1990s was that Sputnik III was originally intended to be the first Sputnik, but that the Russians became more cautious and decided to go with a simpler satellite first.

The immensity of Sputnik III triggered a new wave of recriminations and calls for action. It was apparent that the Soviet Union was preparing for manned spaceflight. Robert Hotz, the editor in chief of *Aviation Week*, described the significance of the launch for the U.S. space community: "Successful launching of the 3000-lb. Soviet Sputnik III should dispel most of the wishful thinking that has hung over the U.S. space policy since the fiery plunge of Sputnik II into the Caribbean [on April 14]. It proves once again that the Soviets' early Sputniks were no lucky accidents. It proves that the

*Leonid I. Sedov, a Soviet academician who is now the voice of the Sputniks, declared, "The new Sputnik . . . could easily carry a man with a stock of food and supplementary equipment."

Soviet space program is a well-organized, consistent effort that is attempting to progress in significant increments rather than simply shooting for some spectacular, international propaganda stunt. It also indicates that the Soviet program has solid and consistent support not subject to the ups and downs of top level policy changes or political whims of the moment. . . . We are still debating in Congress the advisability of establishing a National Aeronautics and Space Agency. We hope Sputnik III will shake some of the Congressional nitpickers out of their lofty perches and prod them into action on this vital message."

In early April von Braun was brought back to Washington to testify in reaction to Sputnik III. This time it was the House of Representatives conducting the hearing, and the main question was how could America overcome the huge Soviet lead in the space race. "Our missile program must be backed by a large budget which permits its steady prosecution over a period of several years," von Braun asserted. "But even with no holds barred, I think it will be well over five years before we can catch up with the Soviets' big rockets because they are not likely to sit idly by in the meantime." He added, "There is a crying need for more money for basic and applied research . . . for development of bigger booster engines." When asked later in the month by a U.S. senator if America was competing with the Soviet Union in space, von Braun replied, "We are competing in spirit only."

On April 17 Generals Medaris and Gavin went to Capitol Hill to testify before the new House Committee on Science and Astronautics to discuss the administration's proposal to create a civilian space agency. Neither man liked the idea. Medaris had been on the road lobbying against it.

Medaris and Gavin addressed what Medaris termed "the problem of space." Gavin, who had resigned on March 31 because of "foot-dragging" in the Pentagon, disclosed that he had sent an emissary to Wernher von Braun in 1957 to tell him to start working on an antisatellite. He believed that the nation was in "mortal danger" because of reports telling him that the USSR would soon have its IGY satellite in orbit. When he was asked if the United States could now shoot down a hostile satellite, he said that it couldn't but should develop the capability as soon as possible.

The two men argued that the nation needed manned military missions, which would use ICBMs to rocket men and supplies to "the scattered and fast-moving battlefields of the future." The argument was that there would soon be a need to send three or four infantrymen in the nose cone of a missile to some remote corner of the Earth with no means of return. Hugh Dryden, director of NACA, testified later in the day and called this idea a "circus stunt" in the same class as shooting a lady out of a cannon.

Despite knowing nothing of what the White House, Air Force, and CIA were working on, the two generals agreed that the United States should develop a spy satellite at an early date. The first formal satellite development program—a photographic surveillance program known as WS-117L—had developed enough security leaks that the Air Force publicly "canceled" it in February 1958. The project did not die but underwent a "covert reactivation" and was renamed CORONA, under joint management of the Air Force and the CIA with extraordinary security.

Work progressed on the CORONA spy satellite, which was so secret that by order of CIA chief Allen Dulles all details were to be passed along verbally and there were to be no documents or written records. In fact, between December 5, 1957, and March 21, 1958, CORONA generated no paperwork. The Air Force Ballistic Missile Division, not the Army Ballistic Missile Agency, worked with the CIA, just as it was doing with the U-2 aircraft. Lockheed, General Electric, and Fairchild were under contract to help build the satellite.

Medaris and von Braun seemed to be assuming—and letting others assume—that space was a military realm and that the Army should and would be the agency to lead America into space.

On April 24 CBS began a two-part television drama on space, titled *Shooting for the Moon*, which made the same assumption. Sponsored by the Chrysler Corporation, it focused on the cooperative roles of the Redstone Arsenal, its civilian prime contractor (which also happened to be Chrysler), and von Braun's Rocket Team in making a first move toward a Moon landing. There was a filmed interview of Medaris by Walter Cronkite.

But lines were being drawn that excluded the Army—there was a line between the military and science, another between military and civilian management of a space agency. A few days after the CBS docudrama, James A. Van Allen, now identified as the man in charge of all U.S. satellite experiments—he was chairman of the Rocket and Satellite Research Panel of the National Academy of Sciences—appeared before the House Space Committee and said that the space program should be placed in civilian hands. "The military will fight it, but the bulk of the scientific community feels as I do." The inexorable transformation of the old civilian NACA to a new civilian space agency had begun in earnest.

Midnight Raids

While the creation of a space agency was a public event, a covert effort was being made to learn more about the Sputniks and their instruments. At one level it was as if the United States needed proof that the Russians actually had the hardware. At another it may have been that sneaking a look at the Sputniks was just too tempting for those Americans who dealt in secrets. The 1958 World's Fair in Brussels was the scene of one audacious intelligence operation.

At the fair the largest of the 162 structures housed the Soviet pavilion, which showed the progress that had been made since the Bolshevik Revolution of 1917. It featured full-scale replicas of Sputniks I, II, and III. (The replica of Sputnik III was added after it went up on May 15.) Russia's exhibit was in a Parthenon-shaped structure with a glass roof supported by cables; a statue of Lenin dominated the interior. Emphasis was on industry, production, and statistical data.

The adjoining U.S. pavilion, on the other hand, was second in size to the Russians' and featured American life and progress in a "relaxed" manner, including a continuous parade of fashion models wearing American-made clothes, and snack bars dispensing hot dogs and hamburgers. The only sober note was a set of three stages displaying three American problems—segregation, inadequate housing, and the waste of natural resources—and the commitment being made to solve them. This was later replaced with a more upbeat display on public health.

As if the contrast between the two pavilions were not enough of a metaphor for America's deficiencies, a caper staged by a motley group of Americans showed how desperate Americans were for intelligence.

Brad Whipple, a Russian linguist on assignment with the 6911th Radio Group Mobile (a field unit of the Air Force Security Service) at Darmstadt, Germany, took leave and headed to the World's Fair with his girlfriend on August 8, 1958.* At the fair Whipple ran into several other Americans in civilian clothes, who recruited him for a "black bag" operation. With orders "not to

*Whipple was part of a top-secret Air Force unit listening to radio transmissions behind the Iron Curtain. He was well aware of Sputnik, as he had been on duty when the Sputnik signal was picked up by a radioman named Malcolm "Hogjaw" Allen. Allen had been monitoring Soviet radio signals while also using a toggle switch to troll low-frequency signals, looking for something more interesting, such as a rebroadcast of the previous day's World Series game. His discernment of the Sputnik signal was likely the first official American contact with the satellite.

get hurt or hurt anybody" and to carry no credentials, a group of men from the CIA, DIA (Defense Intelligence Agency), Navy, Air Force, and other groups broke into the Soviet pavilion and hauled the replica of Sputnik I down from the ceiling, where it was suspended approximately three and a half stories above the floor. For three hours in the middle of the night, the Sputniks were prodded and photographed as if they were the real thing. Today Whipple looks back and recalls that it had all the earmarks of a "college prank, a joke, something that fraternity house boys would try to pull off."

The efforts expended by the intelligence community to learn more about Sputnik, Lunik, and other early Soviet spacecraft included kidnapping satellites. When the Russians showed off the latest in spacecraft in New York City in 1958, a group from the CIA's Technical Services Division (TSD) got into the building, grabbed the Sputniks and had them photographed at a remote location in Long Island, and returned them before dawn. "We went up and back by bus just like kids on a school trip," recalled a participant, who added that the group mustered in Washington, D.C., at a location adjoining the Lincoln Memorial.

This story remained a rumor until 1972, when former CIA employee Patrick J. McGarvey's critical book *C.I.A.: The Myth & the Madness* appeared. "The Sputnik display was stolen for three hours by a CIA team which completely dismantled it, took samples of its structure, photographed it, reassembled it, and returned it to its original place undetected," wrote McGarvey. He did not say that this theft took place in New York City but told the Associated Press that the location of the "birdnapping" was mentioned in one of the hundred or so lines that the CIA had cut out of the book when he submitted his manuscript for review. (Review by the CIA was required under the secrecy agreement he signed upon joining the agency.) One line that was not deleted had to do with the Technical Services Division bragging about the event: "One of their feats, often spoken about at training sessions at CIA, was the stealing of the Soviet Sputnik."

Others had different means of obtaining information. In stunning contrast to these covert operations was a 1957 incident reported in 2000 by physicist Bob Park. He wrote that radio astronomer Herb Friedman of the Naval Research Laboratory was at a conference in Leningrad when Sputnik II was launched: "When he arrived, he asked his Russian hosts what instruments were on board. They did better than that. They took him to a trade fair on the outskirts of the city where a cutaway replica of Sputnik II was on display. He was allowed to take a closer look: a tape recorder, a radio transmitter, and two glass Geiger tubes."

Friedman took down the numbers and the next day went into a store selling scientific equipment for schools and bought two of the instruments, for which he paid a ruble and which he brought home wrapped in a copy of *Pravda*. Park reported what happened next: "Shortly after his return, he was visited by someone from naval intelligence, who was interested in anything Herb might have heard about what kind of scientific data Sputnik II was sending back. So far, he said, the CIA had learned nothing. Herb reached into his coat pocket and took out the two Geiger tubes, still wrapped in a copy of *Pravda*. The astonished agent asked if he could borrow them. Several months later, Herb had just about forgotten the whole episode when an armed security courier showed up at his office to deliver a package marked TOP SECRET."

Sputnik II could have made the same discovery as Van Allen if the Geiger tubes had been able to send back data—Van Allen himself later wrote that the tubes "operated properly and yielded data for seven days." But the Soviet scientists failed to develop the data necessary to interpret the discovery because of a faulty tape recorder, according to Park. A request had been made to the Kremlin to fix the recorder, but that would have meant a launch delay. The request was denied. Van Allen's tape recorder worked. In fact, America would soon use another tape recorder in space to great advantage.

CHAPTER NINE

Ike Scores

The Sputnik mission set off a search for nationhood in the United States even as it seemed to confirm it in the Soviet Union.

—Kevin Klose, *Washington Post*, October 4, 1987

Between early April 1958 and the end of the IGY on December 31, the United States continued to try to rebound from the first two Sputniks. The flights of Explorer and Vanguard, and the discovery of the Van Allen belts, gave America a needed boost; but the nation's space program still needed to be refined to give it direction and to ensure that military wrangling did not jeopardize future success.

In a speech to a joint session of Congress on April 2, 1958, President Eisenhower called for a civilian National Aeronautics and Space Administration (NASA) to be created, with the long-established National Advisory Committee for Aeronautics (NACA) as its core. On April 14, Lyndon Johnson and New Hampshire Republican Styles Bridges introduced the Senate version of the NASA bill (S-3609), and Majority Leader John McCormack of Massachusetts introduced the House version (HR-11881). After hearings, the National Aeronautics and Space Act became law, with the new agency to come into being by October 1. On August 8, Eisenhower nominated T. Keith Glennan, president of the Case Institute of Technology in Cleveland, to be NASA's first administrator. He also nominated NACA director Hugh Dryden a deputy administrator. The Senate confirmed the nominations with little debate on August 14.

The very next day—some six weeks before NASA went into operation—an extraordinary thing took place. A group led by von Braun, in a moment of genuine public petulance, told the nation that they wanted no part in the new NASA. They threatened to leave and go to work in private industry if the transfer occurred. The story was fed to the Associated Press; it ran in newspapers around the nation the next day under headlines like "Protest Led by Von Braun—Scientist Warns of Breaking Up of Missile Team."

As the Rocket Team stewed, the space agency formed quickly. On August 19, the Department of Defense and NASA agreed to transfer nonmilitary space projects to NASA, but they deferred the actual transfers until after the new agency was in place. Glennan and Dryden were sworn in on August 20, and Glennan proclaimed NASA ready to succeed NACA on September 25.

It was clear, however, that NASA was going to have its hands full in acquiring the Huntsville team. The real problem was not as much civilian control as the fact that Glennan did not initially want the whole team for NASA. Glennan visited Huntsville on September 17, 1958, looking to smooth the way to winning the team, but he failed.

On October 1, 1958, NASA officially opened for business, absorbing some 8,000 NACA personnel and a group of laboratories. The headquarters staff gathered at their temporary quarters, the historic Dolly Madison House on Lafayette Square across from the White House, to hear Glennan proclaim the end of NACA and the beginning of NASA. On the same day, by executive order, the president transferred to NASA an assortment of space projects, including Project Vanguard and its 150-person staff from the Naval Research Laboratory; lunar probes from the Army; rocket engine programs from the Air Force; and a total of over $100 million of unexpended funds.

NASA immediately delegated operational control of these projects back to the Pentagon while it put its own house in order. There followed an intense two-year period of organizing, building up, planning, and catching up. Only one week after NASA was formed, Administrator Glennan gave the go-ahead to Project Mercury, America's first manned spaceflight program. The Space Task Group, headed by Robert R. Gilruth, was established at Langley Research Center in Hampton, Virginia, to get this job started.

The new programs brought into the organization were integrated into the NACA nucleus. Many space-minded specialists were drawn into NASA, attracted by exciting new ideas. But NASA was not yet complete; it

T. Keith Glennan, first administrator of NASA, and his deputy, Hugh L. Dryden, being sworn in by President Eisenhower on August 19, 1958 (left to right: Dryden, Eisenhower, and Glennan). (NASA)

needed two components that the Army did not want to give up: JPL and ABMA. On October 14, 1958, Glennan met with Army secretary Wilbur Brucker and two of his top generals to discuss acquiring JPL and a key portion of the ABMA—2,100 individuals in all. Brucker became irate with the idea of losing the von Braun team. Glennan later wrote that he had "not realized how much of a pet of the Army von Braun and his operation had become. He was its one avenue to fame in the space business."*

The morning following the meeting and for the next several days a leaked version of the meeting appeared in the press. The Army came out swinging and found a friendly ear in the press. Depending on what account you read, the proposed transfer was either a "grab" or "a raid." Medaris told the AP that he was on his way to Washington to demand a meeting with the president. He did not get to the president, but he did get to the podium at a meeting of the Association of the U.S. Army, where he declared that this was no time for the Army to surrender its top brainpower to a civilian space agency. Breaking a Pentagon restriction on such talk, Medaris said, "There is a time to paddle and a time to rock the boat."

Glennan was angry but refused to comment in the press. Eisenhower stepped in and gave NASA the JPL, but for the moment at least, he allowed the von Braun team to remain as part of the Army. Glennan had won half a loaf.

*In his unedited diary Glennan made the following comment in a footnote: "Brucker was a man completely dedicated to the Army. In many ways, he was one of the most stupid persons I have ever met. But he fought for his people and they believed in him."

As the infant NASA moved ahead, long-range planning was accelerated. Short-range was another thing. The Advanced Research Projects Agency (ARPA) had been created on February 7, 1958, to ensure that the United States would never again be left behind in the area of new technology. ARPA began with the Department of Defense and carried the ball until NASA was up and running. Eisenhower's point man on ARPA was General Andrew J. Goodpaster Jr., who later recalled: "We simply had to get above this very difficult situation involving the services. You see, if you were dependent on the services that meant you were going to be affected and afflicted by this rivalry rather than having an agency which could go at the problem from a national point of view."

ARPA's interim plans for space exploration were soon approved by the president, and in a sense ARPA was the first U.S. space agency, even though that role was short-lived. For the next year and a half America was kept in the race by ARPA, whose role in space was all-important during those months when NASA plans were tied up in red tape.

As NASA was taking shape, the first American attempts to reach the Moon with unmanned probes were getting under way. The first three attempts were made by the Air Force. On August 17 a Thor-Able rocket without a project name was fired at the Moon and blew up seventy-seven seconds into flight. The Pioneer I launch of October 11 was intended to circle the Moon. It never reached the Moon but did reach an altitude of 79,193 miles—thirty times the altitude of any previous man-made object. Pioneer II, another Lunar probe, went up on November 8, reaching an altitude of only 963 miles before its third stage failed. It was the Air Force's third failure.

Then it was the Army's turn. On December 5 a Pioneer III, launched aboard a Juno II, reached an altitude of 63,580 miles and crashed back to Earth in an uninhabited area of French Equatorial Africa. The only consolation was that it discovered that the Van Allen radiation belt was actually two belts.

At this point, the lunar program was hardly impressive: Of the four attempts to reach the Moon, not one had succeeded.

Project SCORE

The failures of the lunar mission made Eisenhower and his closest advisers even more dissatisfied with the furor being fostered by the military and the large defense contractors. In what would later be seen

Juno II launch, 1959. (NASA)

as one of the greatest and most closely guarded secrets of all time, a group of eighty-eight individuals, including the president of the United States, set about implementing their own plan for a masterstroke in space that was so stunningly executed that Deane Davis, one of those eighty-eight, said "it was done in a manner that James Bond would have been proud of indeed."

It was called Project SCORE (Signal Communications by Orbiting Relay Equipment) and had begun in July 1958 when Roy Johnson, director of the Defense Department's new Advanced Research Projects Agency, paid a visit to Convair Astronautics in San Diego, where the Atlas rocket, America's first ICBM, was being fabricated. The visit came in the wake of Russian success with Sputnik III; Johnson was looking for an appropriate and dramatic response. "We've got to get something big up," Johnson told Atlas program director Jim Dempsey, who replied, "Well, we could put the whole Atlas in orbit." Johnson understood immediately—instead of fabricating a separate satellite vehicle, the rocket itself would be instrumented and go into Earth orbit with its own payload. In short, the rocket would become a satellite when it went into orbit.

Johnson flew back to Washington with the orbiting Atlas-as-satellite idea. A short time later Dempsey and a few other key individuals from Convair came aboard, including Kraft Ehricke, a V-2 alumnus who had first determined that the Atlas was powerful enough to be a satellite without any other stages. Johnson took charge of Atlas 10-B at this point, essentially "buying it" for ARPA, which decided to make it a "talking satellite."

A small team at the U.S. Army Signal Research and Development Laboratory at Fort Monmouth, New Jersey, was brought into the fold. It was the Signal Corps that had been working on space as a vehicle for communications for more than a decade. Project Diana, which had begun on January 10, 1946, was a successful effort to bounce radio echoes off the Moon, 240,000 miles away; technicians at the Monmouth Laboratory had proved for the first time that they could communicate electronically through outer space.

The Signal Corps had tried unsuccessfully to sell the idea of communications satellites to Medaris and von Braun; according to Brigadier General Sam Brown, who was then head of the laboratory, the two men decided orbital communications did not have "enough sex appeal to impress the nation and congress." Medaris was much more interested in spy satellites and schemes that would speed combat troops around the world in the nose cones of rockets.

In early July, General Brown was told by ARPA to construct a communications satellite with a maximum weight of 150 pounds to piggyback the Atlas booster. It was to be designed to test the feasibility of transmitting messages through the upper atmosphere from one ground station to one or more other ground stations. The orbit was expected to be low; this meant the life expectancy of the satellite was only two to three weeks so it could be powered by batteries. Its low orbit would limit opportunities for real-time relay between ground stations; therefore a store and forward mode was added by including tape recorders that could receive a message over one part of the Earth and retransmit it over another. This would also give the satellite a worldwide broadcast capability. Since reliability was a major concern, backup recorders were added to the package.

The idea of putting an Atlas in orbit appealed to Eisenhower as both a show of force and a test of the ability of America to protect a secret. The president personally demanded the strictest possible security. Determined to exert absolute control over leaks, rumors, and second guesses, he laid down one simple warning: If word of this project got out, heaven help those in charge. Ike's man on the job here—and the one with the most to lose—was General Curtis E. LeMay.

Eisenhower also put out the word to the rest of the eighty-eight that if there was a leak he would cancel the project. Period. This frugal, proud, and practical man was threatening to end a major American comeback attempt if his terms could not be met. He felt that America could not suffer any more loss of face. Secrecy was central to the plan. He wanted it to have a low profile so that if it blew up on the launchpad, it would go down merely as an Atlas test failure rather than as another American space debacle. If it failed to go into orbit, it would be just another missile firing.

Project SCORE could also be seen as a means of giving control back to the White House and the Department of Defense. In contrast to the Kremlin, which had total control over its program, Washington seemed to have none. Huntsville was by now calling itself the "Space Capital" and seemed to think it was in charge. When von Braun did not get his way, he would call a friendly reporter and complain. Even American generals had complained bitterly in public when they disagreed with administration decisions. General Gavin's book *War and Peace in the Space Age*, which came out in June 1958, turned out to be, in part, a strident attack on the president and the secretary of defense.

SCORE tested the ability of the system to keep a secret, essential if the United States were to create a system of covert spy satellites. Every one of the eighty-eight in uniform knew that they would be court-martialed if they leaked information. The Army Signal Corps SCORE team that developed "the package" for the project worked in the strictest secrecy. When they were nearing completion in September, all but a few of them—namely, those who would actually sneak the communications equipment onto the rocket at Cape Canaveral—were told that the satellite project they had been working on had been canceled and to forget all about it. General Sam Brown and crew continued working on the payload with the same deadline and were told that it would be tested in aircraft.

Red herrings were placed, false rumors were started, and deception became the order of the day. Cape Canaveral security was told that equipment stowed in a trailer should be left alone because it belonged to the CIA and that the few guys working there (one of whom was Brown) were CIA agents. In fact, the CIA was kept out of the operation completely.

Meetings between key individuals were held in parking lots, drive-ins, and airport waiting rooms. Deane Davis recalled that the final configuration of the 10-B's equipment array was arrived at in a run-down drive-in at the corner of Airport Avenue and Imperial Highway near the Los Angeles International Airport. At an Army base in Arizona, where a SCORE ground station was being set up, the commanding officer threatened to have Signal Corps technicians evicted from the base by the military police unless he was told what was going on. At Cape Canaveral, of the hundreds involved in the launch, less than ten were in on the secret. Even the man in charge of the launch was in the dark, having been led to believe it was nothing more than a routine military missile test.

As SCORE was being set up in the fall of 1958, things were not going well for America's fledgling space program and morale was low. The week of September 20 was catastrophic.

Just after noon on September 24, a Navy Polaris missile was launched from the cape. It went off well but then suddenly shot out of control. The range control officer looked at his radar and realized that the Polaris was not heading downrange over the ocean but toward a densely populated area on the mainland. He had no choice but to punch the destruct button, and, at about a mile above ground, the missile exploded. The first stage broke into a thousand pieces, which rained down on the cape; the second stage splashed down in large chunks in the Banana River near the Hitching Post Trailer Park. Spectators who had flocked to the beaches to watch the announced launch panicked and ran away from or into the Atlantic Ocean. Luckily there were no casualties. It was later learned that the Polaris rocket went out of control when a malfunction developed in the programmer, a small electronic device that directed the control system.

That same week an X-10 pilotless aircraft, which had been sent aloft with a Bomarc interceptor rocket, exploded on landing. Then on September 26 a Vanguard was launched, but the Navy soon announced that the satellite never achieved its proper orbit and, after one or two low orbits, burned up in the atmosphere.

Also that week, Eisenhower suffered his own setback. Sherman Adams, chief of staff and assistant to the president, was accused of accepting gifts and then interceding with federal agencies on behalf of Boston industrialist Bernard Goldfine. Angered by the charges, which seemed to him to be partisan, Eisenhower insisted that Adams was a man of integrity but was forced to ask him to resign because of pressure from Republican members of Congress afraid of losing their seats in the upcoming election.

Lunar Quest

As it turned out, neither the new space agency, nor the announcement of the first manned space program, nor the resignation of Sherman Adams, nor anything else seemed to be able to help the Republicans at the polls. The congressional election saw the greatest Democratic gains since the New Deal: Democrats gained thirteen Senate seats to take a 62–34 majority, and in the House they won forty-seven new seats to take control 282–153. In state elections the Democrats made a net gain of five governorships to control thirty-four state houses against the Republicans' fourteen.

Eisenhower had lost his electorate and the Republican Party was in disarray, but the president was determined to make the best of his nascent plans for space. On November 17 Lyndon Johnson spoke out in support of the U.S. proposal to create the Ad Hoc Committee on the Peaceful Uses of Outer Space at the United Nations General Assembly in New York. Eisenhower sent a plane to Texas to fly Johnson to New York, thus demonstrating bipartisan support in the United States for international cooperation.

By December 1958, the Army's SCORE satellite was ready to be launched. The 150-pound payload was attached to the side of an empty Atlas rocket casing. The missile tracking system was removed and its guidance system changed. A prerecorded message prepared by a member of the SCORE team at the cape was loaded in the tape recorders. Herbert Hawkins of the Signal Corps, chosen because of his strong, pleasant voice, recorded a few nonpolitical lines from an American patriotic document hours before the scheduled launch.*

At the last minute, however, President Eisenhower himself, now convinced that the mission would succeed and pleased that the extreme security precautions had worked, decided at the urging of Roy Johnson to record a Christmas message to the world. The president's tape was rushed to the Cape Canaveral launch site. It arrived after the communications package had been sealed and bolted to the Atlas missile, which was on the launchpad and fueled. So on the morning of December 18, the Signal Corps transmitted the president's message across Cape Canaveral to the communications payload on the waiting rocket. Ike's message was recorded on both the primary and backup tape recorders.

That night, a four-ton Air Force Atlas ICBM was fired into orbit from Cape Canaveral, making it by far the largest object yet known to circle the globe.† Most of those in the blockhouse had no idea what was going on until the

*In a 1981 article on SCORE, General Brown noted that "nobody seems to recall any longer what [the original message] actually was."

†The core stage of the R-7 for the first three Sputniks did in fact go into orbit but this was not known at the time. These objects weighed 12,000 pounds but this fact was a long-held Soviet secret.

missile had broken out of its normal trajectory and headed for orbit. At one point the range safety officer saw it go off his plotting board and reached for the destruct button. "Don't worry about it, son," said General Leighton Davis, one of the eighty-eight. "I'll assume responsibility. Let it go."

America had put an object into orbit with a 150-pound payload. At 8,750 pounds, the orbiting rocket-satellite was three times heavier than Sputnik III; America was now ahead in the Cold War weight-lifting contest. It was a great surprise, especially to the growing corps of reporters covering the new beat of space who knew nothing of SCORE at the time it was launched. But the bigger surprise was yet to come. At 3:15 P.M. on December 19, an interrogation signal from Cape Canaveral successfully triggered the broadcast of SCORE's previously recorded and stored Christmas message from President Dwight D. Eisenhower.*

Immediately and for the next thirteen days the following tape-recorded message was heard intermittently throughout the world:

> *This is the president of the United States speaking. Through the marvels of scientific advance, my voice is coming to you from a satellite circling in outer space. My message is a simple one. Through this unique means I will convey to you and to all mankind America's wish for peace on Earth and goodwill toward men everywhere.*

The tape recorder continued to work perfectly, responding to seventy-eight real-time store and forward voice and teletype transmissions between ground stations. The SCORE satellite was the first voice relay satellite sent into space and proved that voice and code could successfully be relayed, stored, and forwarded over tremendous distances. The versatility of the equipment was demonstrated time and again. For instance, the first version of Ike's message had picked up some static, so on the third pass over the cape a signal was sent that created a clean, clear backup version.

The White House, the Air Force, the Army Signal Corps, and the new agency ARPA had worked together successfully to get America back in the space race. The Air Force provided the vehicle and the Signal Corps the payload. Eisenhower was in high spirits when he brought clueless reporters into press secretary James C. Hagerty's office to hear his greeting from space. "He took Mr. Hagerty's chair and sat hunched over it, listening to his

*On the first orbit, as the satellite passed over California, the primary recorder did not respond properly.

own voice," reported the *New York Times*. "The President called the demonstration one of the astounding things in this age of invention."

This was a White House story, not one from Huntsville or Cape Canaveral, and Ike was in total command of the situation. Medaris, von Braun, Hagen, members of Congress, the media, the scientific community, Lyndon B. Johnson and the rest of Congress, the leaders of the IGY, and almost all the other individuals and institutions in the space business were conspicuously absent. Although he was on the list of eighty-eight, even Secretary Brucker of the Army was kept out of the loop in the final days and was furious when he first got word of the launch from a reporter.

The Huntsville contingent was uncharacteristically silent in public and privately dismayed by a major space event that had, by design, excluded them. Even if just for a moment, Ike had silenced and humbled his severest critic, depriving von Braun of another chance for self-promotion. (Earlier in the year Ike had told senior aides that he was "getting a little weary of von Braun's publicity seeking.") In fact, von Braun would not acknowledge SCORE in his encyclopedic 1966 *History of Rocketry and Space Travel* and instead termed the Echo satellite orbited in 1960 "the first" communications satellite.

Those who were not mad at being excluded were incredulous. Richard H. Jackson, who worked at the Cape at the time and would later work on the Atlas program, was on his way home driving through Cocoa Beach when he heard the news on the radio. "I couldn't believe my ears. An Atlas in orbit. This had to be a mistake."

Heard around the world, Eisenhower's personal message of peace gained remarkably good international press. American journalists in Moscow reported that average Russians, rather than Russian officials, were offering congratulations.

The press, which had been totally excluded from this story, gave SCORE mixed reviews. Doug Dederer wrote in the *Cocoa Tribune* that it "restores the Free World's faith in U.S. missile abilities," and *Newsweek* said that in one moment the "missile gap" had been closed. Under the headline "4½ Ton Satellite Shows What U.S. Can Do," *U.S. News & World Report* asserted that America had taken a "lengthening lead in the space race," and *Life* termed it "America's unique peaceful Christmas gift to itself and the world." The *New Republic*, on the other hand, called it "an extravagant publicity stunt" and argued that it wasted money needed for scientific research. Years later, an official Air Force history of the Atlas Ballistic Missile Program dismissed it as a "public relations effort."

The conspirators relished their role in this spectacular deception. Roy

Johnson awarded each of them a certificate granting life membership in Club "88," composed of "trusty SCORE Crows" who were "duly initiated into solemn mysteries of the Society for the Correction of Russian Excesses (SCORE)."

America now had an object in orbit bright enough to be seen with the unaided eye, appearing as bright as the stars in the Big Dipper. SCORE had cost ARPA a mere $700,000 over and above costs already absorbed by the Atlas program and had begun to pull America out of its Sputnik gloom. The alien beep of Sputnik had given way to the reassuring avuncular midwestern tones of Eisenhower; suddenly there was an entrepreneurial interest in space as a platform for forwarding messages of all sorts from one point to any other on the planet. Many, especially in the capitalist countries, began to see the Earth-circling satellite as a vehicle for making money rather than making war. Hans K. Ziegler, writing in 1960 when he was chief scientist of U.S. Army Signal Research and Development Laboratory, characterized SCORE as the first prototype of a communications satellite and the first test of any satellite for direct practical applications.

SCORE also was a powerful display of missilery. "The iron fist was always there under the velvet glove of the Eisenhower Christmas message," Deane Davis later wrote. "Even 10-B if you'll recall was made of steel." For Ike and others close to him, not only was SCORE a test run to see how closely other secrets could be kept for America's first generation of spy satellites, but it demonstrated that the United States saw space as a great platform for gathering military intelligence—that is, secret military intelligence.

The last time that Eisenhower's voice could be heard from space was on December 31, the final day of the International Geophysical Year. During the thirteen-day life of its batteries, the satellite was interrogated by Signal Corps ground stations seventy-eight times, using voice and teletype messages for the communications tests with excellent results. On January 21, 1959, the heaviest object to have been launched into orbit and the first communications satellite reentered the Earth's atmosphere and was lost somewhere in the African highlands.

Von Braun's Rocket Team Moves to NASA

At this point, von Braun, under ARPA sponsorship, was given the job of designing a massive rocket called Saturn. The Defense Department, however, as ARPA's parent organization, was coming to the conclusion that it did not need such a superbooster for missiles and was beginning to withdraw support over the objections of both ARPA and ABMA. By the process of elimination, the Saturn rocket was beginning to look as if it could and should be the workhorse of manned spaceflight, and von Braun and company knew that they were going to have to go to NASA. Glennan decided to take the whole team. On October 21, 1959, Eisenhower signed an executive order indicating that personnel from the Development Operations Division of ABMA would be transferred to NASA, subject to the approval of Congress.

The debate continued, however, as some of Eisenhower's critics were concerned about the potential transfer of what they believed to be a military function to a civilian agency. In the end, both the secretary of defense and President Eisenhower concluded that the entity ABMA and the Saturn Project should be transferred to NASA where it belonged. A new home was found for the von Braun team by setting aside a complex within the borders of Redstone Arsenal, where they had been all along.

The transfer was anything but smooth. Medaris quit (insisting that by this point he was more "tired than ired"), and, once again, von Braun threatened to, but backed away when he realized that he was being given the bulk of the space agency's space booster systems work and the responsibility for launch operations. As part of the bargain, the team and its leadership would be held together as the NASA Rocket Team. The transfer included von Braun and 2,327 other rocket and satellite specialists. NASA established the George C. Marshall Space Flight Center at Redstone Arsenal in the spring of 1960.

On March 15, the White House announced that the George C. Marshall Space Flight Center would assume responsibility for providing the launch vehicles for civilian exploration of outer space. The problem of naming the civilian space center after a military man was obviated by Marshall's other distinctions—he was a secretary of state, a Nobel Peace laureate, and initiator of the Marshall Plan. When the name was first suggested by NASA administrator T. Keith Glennan, Eisenhower responded, "I can think of no one whom I would more wish to honor."*

The Army had lost its position in the space program. Although Medaris would not admit it at the time, this was the reason he quit. On the twenty-fifth anniversary of the founding of the Marshall Space Center, he told the *Birmingham News*: "I don't feel too inclined to celebrate the birth of NASA.

*The center was activated on July 1, 1960; the formal dedication ceremonies were held on September 8. Ike and von Braun seemed to have finally come to a truce. Von Braun proudly showed the president his monster booster rockets.

I retired because of it—I wasn't about to sit there and watch one of the finest commands ever put in place torn up . . . so I retired."

Von Braun dominated NASA. He was uniquely able to get things done. During the late 1950s and early 1960s, he delivered more than one speech a week—often saying that *everyone* was a winner if America committed to a major space program. In January 1959 he told the Associated General Contractors of America that they would be building roads on the Moon in twenty years. Not only did he motivate with promises; he would sprinkle in threats: At the same convention he told the contractors that the Russians were still very much ahead of the United States and that unless Americans began to move faster, they might find themselves "surrounded by several planets flying the hammer and sickle flag." Later in the year, at a press conference, he said provocatively: "I am quite sure that when our young astronauts blast off into space they will meet Russian customs officers."

But even more important than von Braun, NASA now had what would become the Moon rocket. Four years before President Kennedy declared the lunar expedition a national mission and a month before the first Sputnik, the Army already was at work on what would become the upward force of the Apollo program. While still preparing for the launch of its first Jupiter, the Rocket Team began studies of a booster ten times more powerful than the Jupiter. The tenfold increase in thrust could put a weather and communications satellite into orbit around the Earth or propel a space probe out of Earth's orbit.

This change of emphasis from intermediate-range and intercontinental ballistic missiles to the super-rocket capable of space exploration signified a change of attitude at the Department of Defense. The Rocket Team had long ago set its sights on the new super-rocket, later to be named Saturn. In December von Braun's group set out arguments for the new booster program, which was approved by ARPA and adopted by NASA. The super-rocket would develop 1.5 million pounds of thrust and serve as a stepping-stone to an even larger rocket pointing its nose toward a trip to the Moon in 1967.

The Lunik Probes

If Khrushchev was impressed with SCORE, he said nothing of it in public, but he did respond with his own gesture nine months later. After the May Sputnik III success, the remainder of 1958 saw no fewer than

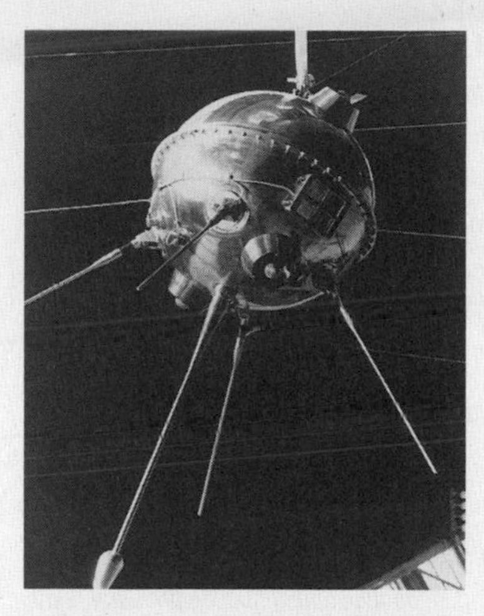

Lunik II model on display at the Cosmos Pavilion for the Exhibition for the Achievements of the National Economy in Moscow. (NASA)

three attempts to send Lunik probes to the Moon, all of which failed. (Although Lunik I missed its target, it was the first object to leave Earth and enter an orbit around the Sun.) On September 13, 1959, the Soviet Lunik II became the first probe to land on the Moon. The following month, Lunik III flew behind the Moon, sending the first images of the far side of the Moon back to Earth. Two days after Lunik II's landing, Khrushchev arrived in the United States for a tour, bringing Lunik II souvenirs for his hosts, including the president.

Luna was the Soviet name for its series of Moon probes, but Americans, enamored of the *-ik* suffix, to this day insist on calling them Lunik.

Desperate for details on Lunik, the CIA was enlisted in a scheme that recently was made public. The official version by Sydney Wesley Finer began: "A number of years ago the Soviet Union toured several countries with an exhibition of its industrial and economic achievements. There were the standard displays of industrial machinery, soft goods, and models of power stations and nuclear equipment. Of greater interest were the apparent models of the Sputnik and Lunik space vehicles." At this point the article mentions that the story about to be retold was of the second look at the Lunik: "U.S. intelligence twice gained extended access to the Lunik, the second time by borrowing it overnight and returning it before the Soviets missed it."

If peeking at hardware in the wee hours seemed to be a sophomoric and essentially harmless response to Sputnik, other actions taken during the IGY's waning days—before the Medaris–von Braun team broke up—were not.

The Nuclear Summer of 1958

The summer after the Sputnik launches, America plunged itself into a binge of bold—some would say reckless—nuclear demonstrations. The United States detonated seventy-seven aboveground tests in 1958—a record eclipsed only in 1962.

Early in the morning of August 1, 1958, a Redstone missile was used in a nuclear test, fired from Johnston Island in the Pacific, 700 miles from Hawaii. At approximately 12:52 A.M. Hawaii time, the first stage of the missile rose to a height of what observers claimed was 100 miles and exploded. The nuclear blast illuminated the night sky first with a bright flash and then with a mushroom cloud clearly visible in Hawaii. Later in the day, the *Birmingham News* carried a banner headline over an Associated Press dispatch, "First Redstone A-Missile Exploded 100 Miles in Space," and a subheadline of "Near Panic Grips Hawaii." It was later reported that the bomb's electromagnetic pulse knocked out most of Hawaii's telephone system. America had now reacted to the Sputnik challenge with audacity, daring, and a clear eye on making space a laboratory for thermonuclear offense and defense. The stated purpose of the first rocket-launched nuclear test by the United States was to study the effect of nuclear explosives at high altitudes.*

The first nuclear rocket detonation was given the code name Teak; another, code-named Orange, was fired on the August 11 and detonated at half the altitude. Little information was given out at first, but it was later disclosed that the Teak exploded about halfway to the 100 miles originally reported, while Orange went off at 24 miles.

According to Teak participant Bernard J. O'Keefe, author of *Nuclear Hostages*, anyone within fifty miles of the blast would experience retinal burns, so the tests had been conducted on Johnston Island, a tiny spit of land just large enough for a runway. It was the only piece of real estate governed by the United States in the whole Pacific Ocean that did not have inhabited islands within its line of sight. "When Teak was detonated, it was a frightening sight," recalled O'Keefe. "Because of the long range of the X-rays in the low density of the atmosphere, the fireball grew very rapidly. In 0.3 second its diameter was eleven miles; in 3.5 seconds it increased to eighteen miles. The fireball also as-

*Von Braun was on the island for the test, while Medaris chose this day to announce the debut of a crimson, single-breasted, three-button blazer with the ABMA rocket logo on the left breast pocket. It was suggested as off-duty wear for the officers of his rocket corps.

cended at a high rate of speed, initially at one mile per second. Surrounding the fireball was a very large red luminous spherical wave. About a minute or so after the detonation, the Teak fireball had risen to a height of ninety miles, clearly visible by line of sight all the way to Hawaii, seven hundred miles away. In addition to blocking radio communications over a radius of several thousand miles, the Teak shot created an 'artificial aurora,' which was observed in the Samoan Islands, two thousand miles from the point of burst."

The effects were far-reaching: "[Teak] blacked out radio communications from and to Sydney, Australia; Wellington, New Zealand; Honolulu; Vancouver, British Columbia; and San Francisco for a period of hours," according to a report by the U.S. National Bureau of Standards. The Washington-based magazine *Science Trends* said that some radio circuits spread out over thousands of square miles were "out for days."

Almost a year later, on July 9, 1959, von Braun returned to Hawaii to address an Association of the U.S. Army audience and revealed some added details of the two detonations: "[They were] the first megaton devices ever detonated in the stratosphere by the United States and the first ever carried aloft in U.S. ballistic missiles."

What was the purpose of these blasts?

Von Braun suggested that one objective was to determine the extent of retinal eye damage caused by exposure to a high-altitude nuclear detonation. "Rabbits were placed at distances up to 300 nautical miles from the point of detonation in order to obtain necessary data," von Braun dryly reported, explaining that the experiment showed that the high-altitude bomb sent out a great percentage of its energy in a fraction of a second after the blast as compared with the energy from a low-altitude detonation, which did not move nearly as fast. The blinking reflex in a rabbit is just over a quarter of a second, which meant the Teak rabbits did not have time to close their eyes before nearly all of the radiant exposure hit and burned their retinas. The burning effect would be the same in humans, whose blink response is just under a quarter of a second—still much too slow to beat the blast.*

There was more to Teak and Orange than proof that a powerful thermonuclear-tipped rocket could cause blindness. Von Braun admitted that

*As for the effect of the blasts on the estimated 2 million birds on Johnston Island, Von Braun responded: "A water spray was employed to simulate rain and this kept the birds grounded. Smoke together with local clouds further attenuated the thermal effects. By and large, the protective measures tied to the countdown were very successful. A few birds were killed or blinded."

another objective was to see to what degree the blasts could disrupt the ionosphere and knock out radio communications. He would go no farther. Newspapers had speculated that the blasts could trigger the premature firing of an incoming ICBM, which was later confirmed. This was, in fact, the first attempt at creating an antimissile defense—a nuclear counterpunch in the form of a ring of hydrogen bombs exploded to encircle North America in a shield of radiation.*

The testing then went higher into space a little more than a month later. On August 27, 1958, at an altitude of 300 miles, the first of three more nuclear weapons were set off in space. Code-named Argus, these tests were kept secret from the public until portions of the story started to leak in the press in 1959 and the *New York Times*, which had prior knowledge of the original story and voluntarily suppressed it, now released it. The *Times* had taken its cue from scientists who, in the paper's own words, "feared that prior announcement of the experiment might lead to protests that would force its cancellation."†

Months after the fact, in March 1959, the Defense Department held a press conference to explain the importance of Argus in its future plans. Argus demonstrated that a "telecommunications mirror" could be created in the ionosphere which would serve as a perfect reflective medium for radar to peer behind the Iron Curtain. The existing ionosphere was an imperfect mirror impaired by solar flares and magnetic storms, but Argus proved that an artificial ionosphere could be created. The plan was to enhance the mirror effect with billions of small copper needles forming a massive reflective belt. The United States planned to add to the number of copper needles if the experiment proved to be successful. This plan to fill the ionosphere with copper needles was strongly opposed by the International Union of Astronomers and others, and was never put into play.

The three Argus bombs were set off above the South Atlantic Ocean near the Falkland Islands from the guided missile ship USS *Norton Sound*,

*The Russians no doubt knew of this idea and may in fact have also tested it, so that either superpower could use it. In other words, it gave the United States no clear defensive advantage.

†In *Without Fear or Favor,* Harrison E. Salisbury, a former foreign correspondent and assistant managing editor at the *Times,* wrote of the holding of the Argus story as an example of journalistic responsibility: "The story was held up not because of pressure or interference by the Government or the White House but because responsible journalists on the *Times* believed this action was in the overall interest of science and the country."

in the part of the lower Van Allen belt closest to the Earth's surface. The man who devised this experiment and sold it to the Pentagon was Nicholas Constantine Christofilos, a barrel-chested maverick with no degree in physics. He immodestly declared, "This was the most fantastic experiment ever conducted by man."*

In a 1959 profile of Christofilos, newsman Edwin Diamond described the magnitude of the tests: "Argus was an audacious and impudent experiment. It used all of the space surrounding the earth as a gigantic laboratory to test a Christofilos idea that electrons born in the nuclear bursts would travel quickly around the earth's magnetic field. . . . The electrons behaved precisely as Christofilos predicted they would—eerily beautiful auroras were created over the Atlantic; short-wave radio facilities, which depend on the ionosphere, were upset; and a thin sheet of man-made radiation blanketed the earth for three weeks."

Its "aurora" was awesome—eyewitnesses compared it to a full Moon. The total effect had been to create what amounted to an artificial ionosphere or a third Van Allen radiation belt. These gigantic experiments in fact created new magnetic radiation belts encompassing almost the whole Earth and injected sufficient electrons and other energetic particles into the ionosphere to cause worldwide effects—essentially a veil of electrons draped around the Earth.

The Pentagon's attempt to keep Argus data secret was foiled, in part, by the power of the blasts themselves. Soviet monitoring stations in locations as far-flung as extreme northwestern Russia recorded the waves clearly and reported this fact in 1960. One of the reasons that the information had been kept secret in the first place was that the United States was testing atomic devices far from its known proving grounds.

Thinking the Unthinkable

At the time of Teak, Argus, and other early tests, Americans were assured by those in control that they posed no hazards to human life, but this was said about other nuclear events that have, over time, proved otherwise. The legacy of using space as a laboratory for the effects of

*Christofilos, allegedly the inspiration for the James Bond movie *Goldeneye,* was a bona fide self-educated genius, who used his own design ideas for atom smashers to land a job at the Livermore Radiation Laboratory.

nuclear weapons is still largely unexplored, unquestioned, and unresolved. Fortunately, other destructive ideas of the period did not come to pass.

For a time, there was talk of exploding an atomic bomb on the Moon as a direct rebuttal to the shock of Sputnik—exactly the thing some Americans guessed the Russians were going to do in the wake of Sputnik to mark the anniversary of the Bolshevik Revolution. The Russians put up a dog instead, but the idea did not die—and Teak and Orange gave it new life. An editorial in the *Birmingham News* the day after Teak heralded the introduction of nuclear weapons into space and speculated: "This kind of blast very conceivably could be an immediate preparation for a blast on the surface of the moon."

At the time of the Birmingham editorial, the United States had actually begun seriously studying the idea, a fact that did not become public until September 1999 when Keay Davidson revealed it in his biography *Carl Sagan: A Life.* In May 2000 one of the participants in the study, Chicago-based physicist Leonard Reiffel, wrote about it in a letter to the British journal *Nature*. He directed the study at the former Armour Research Foundation, now part of the Illinois Institute of Technology under a contract from the Air Force's special weapons center in Albuquerque.

The purpose envisioned for such an explosion was to make a showy display, clearly visible from Earth, demonstrating that the United States controlled the "high ground" of space and that American missiles were as daunting as their Soviet counterparts. But as Reiffel, at age seventy-two, told the *New York Times* in a May 16, 2000, interview, "The foremost intent was to impress the world with the prowess of the United States. It was a P.R. device, without question, in the minds of the people from the Air Force."

Of course, the secret project, titled "A Study of Lunar Research Flights," was never carried out; but its planning included calculations by astronomer Carl Sagan, then a young graduate student, of the behavior of the dust and gas generated by the blast. Under the scenario, a missile carrying a small nuclear device was to be launched from an undisclosed location and travel 238,000 miles to the Moon, where it would be detonated upon impact. The planners decided it would have to be a Hiroshima-size atom bomb because a hydrogen bomb would have been too heavy for the missile. Since the Moon lacked the atmosphere of the Earth, there would be no mushroom-shaped cloud there; instead, dust from the lunar crater would shoot out in all directions. As of this writing, there has been no official Air Force explanation of why the project was abandoned.

* * *

If America had its secrets, so did Russia. Shrouded in mystery for more than two decades was an early September 1957 volcanolike explosion of nuclear waste buried in underground vaults in the Ural Mountains. Information on this did not surface until November 1976, when an article appeared in the British weekly *New Scientist* by Zhores Medvedev, an exiled Soviet biochemist then working at the National Institute for Medical Research in London. He disclosed that hundreds of people were killed immediately and thousands more were severely contaminated. He added that tens of thousands were affected. When Medvedev came out of the Soviet Union, he was appalled to learn that the rest of the world knew nothing of this calamity.

A follow-up study by the Oak Ridge National Laboratory in late 1979 not only confirmed Medvedev's report but added some chilling details. For instance, a comparison of maps of the southern Urals before and after the accident showed that more than thirty names of small communities with populations under 2,000 had been deleted from the later maps, indicating a considerable movement of people from the area in question. The Oak Ridge scientists also determined that an elaborate system of canals was built to bypass fifteen miles of river valley below the site.

In 1984 the site was photographed by astronauts aboard the space shuttle Challenger so that scientists might effectively determine the residual radiation damage. The U.S. government already had pictures of the affected zone. They were not shared with the public because they carried a "top-secret" stamp and would have been a tip-off as to the quality of U.S. satellite reconnaissance photography.

The United States' Eye in the Sky

The CORONA program was developed at a moment when it was still extraordinarily difficult for the United States to gather information by any means other than by U-2 aircraft flying over "denied areas" in the Soviet Union and China. The U-2 went into operation in the summer of 1956 and was expected to have a relatively short life because America believed that the Soviets would soon have radar sophisticated enough to track the aircraft. American intelligence had misjudged the Soviet technology, because on the very first U-2 flight over Soviet territory, their radar promptly picked it up and continuously tracked it. The Soviets registered a formal protest within days of the incident, and the United States decided to restrict flights over the Soviet Union. Over the next four years sporadic flights were made without incident, although they usually were traced. But on May 1, 1960, the Soviets shot down an American U-2 aircraft and captured its CIA pilot, Francis Gary Powers.

This dramatically increased the urgency of finding an alternative way to gather intelligence. CORONA was created in deepest secrecy by Lockheed Corporation employees working in rented quarters at a helicopter manufacturing plant in Menlo Park, California. CORONA was designed to take pictures in space of the Iron Curtain countries. While in orbit, it would photograph with a rotating stereoscopic camera system and load the exposed photographic film into recovery vehicles, known as "buckets." The buckets would be thrown out of orbit, float down on parachutes, and be recovered in midair by Air Force C-119 aircraft.

From January 1959 through July 1960, thirteen CORONA launches were failures—although some came close. For instance, on April 13, 1959, Discoverer II was the first CORONA satellite to go into orbit and later eject a film capsule. But the capsule went into reentry at the wrong time and landed on the snowy island of Spitsbergen, north of Norway in the Norwegian archipelago, instead of hitting its intended target near Hawaii. Air Force search planes never found the capsule. However, a Soviet icebreaker was spotted in the area, and the CIA suspected the capsule might have ended up in Moscow.

"The first partial success came August 10, 1960, when the capsule of Discoverer XIII was recovered from the Pacific Ocean, the first time an object had been retrieved from space," wrote Mike Langberg of Knight-Ridder Newspapers in 1995, one of the first journalists to write about CORONA. "The Air Force, eager for publicity, touted the accomplishment, although Discoverer XIII was only a test flight carrying an American flag as its primary cargo. Nervous CORONA technicians barely had time to extract a secret 10-pound device, intended to measure whether Soviet radar could detect the orbiting satellite, before the capsule went on a press tour culminating in a ceremonial presentation to Eisenhower at the White House."

(From this point forward the CORONA was hidden behind "cover" missions. The CIA's Richard M. Bissell Jr. contrived the cover story that mice and monkeys were sent aloft, hailing the satellites' mission as "biomedical and technological experiments." For many years the mission was so secret that even the name was kept from the public. The first the public knew of CORONA officially was in February 1995, when President Clinton issued an executive order directing the declassification of images acquired by the first generation of

U.S. photo-reconnaissance satellites. The order provided for the declassification of more than 860,000 images collected between 1960 and 1972.)

On August 18—109 days after the U-2 was shot down—the United States launched its first successful reconnaissance satellite. Discoverer XIV dropped a canister of exposed film from space. It was recovered in midair by an Air Force cargo plane. The eighty-four-pound gold-plated drum, suspended beneath a parachute, was retrieved near Hawaii. Inside the capsule were twenty pounds of film showing pictures of Soviet military installations. After seven complete passes over the Soviet Union, this one spacecraft had provided more photographic coverage of the Soviet Union than all of the previous U-2 missions combined.

The "missile gap" between the United States and the USSR was front-page news, and a major issue in the 1960 election campaign then in full swing. Military and political leaders fretted about America's second-place status in the nuclear-arms race. CORONA later revealed that the Soviet Union did not have 3,000 long-range missiles, as some U.S. analysts feared. Neither were the Soviets turning them out "like sausages," as Nikita Khrushchev had warned. They had only a half dozen.

Between 1959 and 1972, CORONA spanned 145 missions that provided intelligence the government deemed virtually invaluable. The CORONA cameras, known by the code name Keyhole, quickly improved. The original KH-1 was replaced by a succession of models, culminating in the KH-4 in 1962. By the end of the decade, the KH-4 could pick out objects six to ten feet across. Under optimal conditions it could spot cars and trucks. By 1970, flights remained in orbit for nineteen days, provided accurate altitude, position, and mapping information, and returned coverage of 8.4 million square miles per mission. After thirteen years, the CORONA program ended with a launch on May 25, 1972, ultimately having delivered images of 750 million square miles of Earth's surface, mostly in the Soviet Union and China.

"I wouldn't want to be quoted on this," President Johnson said of CORONA, "but we've spent a billion dollars on the space program. And if nothing else had come out of it except the knowledge we've gained from space photography, it would be worth 10 times what the whole program has cost. Because tonight we know how many missiles the enemy has and, it turned out, our guesses were way off. We were doing things we didn't need to do. We were building things we didn't need to build. We were harboring fears we didn't need to harbor."

If CORONA was born out of anxiety and desperation, it offered relief from the same. Satellite surveillance gave the United States confidence to negotiate nuclear-arms-limitation treaties with the Soviet Union in the 1970s because of the ability to watch Soviet weapons production. Indeed, it was argued that U.S. negotiators often knew more than their Soviet counterparts about the exact contents of the Soviet arsenal, although America never admitted to using satellites for arms-control treaty verification.

Race to the Moon

In July 1958, the Advanced Research Projects Agency (ARPA) announced its intention to support the development of a super-rocket powerful enough for a lunar launch, and a month later the Saturn rocket was selected.

During the presidential election year of 1960, NASA got its mission and was given the Rocket Team and Saturn to carry it out. NASA's first ten-year plan was submitted to Congress in February of 1960 and called for a broad program of nearly 260 varied launches during the next decade, including Earth orbital satellites, lunar and planetary probes, and manned flights to orbit Earth and fly around—but not land on—the Moon. The costs, estimated by guesswork and intuition, were given as $1–1.5 billion per year. The House Committee on Science and Astronautics considered it a good program except that it did not move ahead fast enough. But there was a question of how quickly the technology could move. The day after the ten-year plan was announced, NASA's first unmanned Mercury-Atlas disintegrated and fell into the ocean fifty-eight seconds after takeoff from Cape Canaveral.

On July 28, 1960, NASA announced a new manned spaceflight program called Apollo. The program would be carried out with a series of large, ever more powerful Saturn rockets, which would become the workhorse of the program.* Initially, NASA's aim was to put three astronauts into sustained Earth orbit and then into a flight around the Moon. But first NASA had to get a man into space. Its immediate goal was the successful orbiting and return of a human aboard a Mercury spacecraft.

By the time of the presidential election of 1960, the worst birth pangs

*Originally termed the Juno V, the super-rocket was renamed Saturn in Huntsville work papers of mid-1958, and the new name received official status in early 1959.

were over for NASA, which, with the help of ARPA and an eager Congress, had become a thriving organization. But after John F. Kennedy was narrowly elected over Richard M. Nixon in November, a mood of uncertainty took hold. Jerome B. Wiesner, the president-elect's science adviser, chaired a committee that produced a report both critical of the space program's progress to date and skeptical of its future. As if to underscore the mood of doubt, on November 21, 1960, the first Mercury-Redstone lifted a fraction of an inch and settled back on its launchpad. The failure set a low point of morale for the engineers working to put the first American into space. What Kennedy understood, however, was that the fledgling space agency had the support of the American people.

The Russians were not idle during this period. On April 12, 1961, Major Yuri Gagarin went into space for an eighty-nine-minute orbital flight in Vostok 1. After one orbit, he reentered the atmosphere and landed safely, admitting him to the elite aeronautical group who were the first to do what was once thought impossible—the American Wright brothers, the American Charles A. Lindbergh, and now this charismatic Russian.

The United States responded on May 5, 1961, when thirty-five-year-old astronaut Alan Shepard's Freedom VII spacecraft carried by a Redstone booster took a fifteen-minute suborbital ride and was picked out of the ocean some 300 miles downrange of the cape. Shepard had undertaken an incredibly dangerous and heroic mission.

Americans did not lack resolve, but their technology was still lagging, especially in the vital imperative of rocket power. By any measure, the United States was still behind. Gagarin had flown around the Earth for a total of 24,800 miles versus Shepard's 300. His Vostok spacecraft weighed 10,428 pounds in orbit, against Mercury's 2,100 pounds in suborbit. Gagarin spent about eighty-nine minutes in a state of weightlessness, Shepard five.

President John Kennedy and his space-oriented vice president, Lyndon Johnson, had won by running against the "drift, delay, and dilution" of the Eisenhower-Nixon administration in dealing with the space challenge. Now the new administration was hard-pressed to fashion a response to Gagarin's feat. Johnson, chairman of the National Aeronautics and Space Council, was ordered by Kennedy to come up with a plan.

Within a matter of days, Johnson came back with the answer: Put an American *on* the Moon before the end of the decade and bring him home to a heroic welcome. It was an answer that served many constituencies. Some saw the quest to land on the Moon as imperialism ("Tomorrow the country that controls the moon will control the earth," according to a vice president

Astronaut Alan B. Shepard, moments before he became the first American in space. (NASA)

of the Glenn L. Martin Company who was quoted in *Aviation Week* on February 10, 1958); others saw it as pure exploration.

Looked at cynically, this quest drew attention from the then-prevalent economic instability and from the recent debacle of the Bay of Pigs invasion of Cuba; even more cynically, it would provide an economic bonanza for Johnson's home state of Texas, where the ground-control center was to be established in Houston. Kennedy saw the Moon landing as something that, with any luck, would occur toward the end of his second administration in 1968, when he hoped to be the president to express congratulations to the first American on the Moon.*

*The timing would be correct, but Nixon, not JFK, would be the chief executive to preside over Kennedy's event. Tragically, on the same weekend as the Moon landing, Senator Edward Kennedy drove his Oldsmobile off Dike Bridge on Chappaquiddick Island, Massachusetts, killing Mary Jo Kopechne, a twenty-eight-year-old Washington secretary. Many Americans did not hear of this until they got their newspapers on Sunday morning, when the two events appeared in stunning juxtaposition. For instance, the first and second headlines in the *Washington Post* read "Apollo in Orbit, Plan for Landing" and "Kennedy Passenger Dies in Car Plunge." On Monday, July 21, when the actual Moon landing was reported, the news was worse. The front-page headline in the *Baltimore Sun* announced "Kennedy Will Face Charges in Leaving Accident Scene," while the *Philadelphia Inquirer* carried the front-page story "Police to File a Complaint on Kennedy."

President John F. Kennedy addressing a joint session of Congress on May 25, 1961, on the goal of landing Americans on the moon before the end of the decade. At the time of the address America had logged 15 minutes and 22 seconds of manned spaceflight while the Russians had clocked 108 minutes. (NASA)

But the proposed Moon-shot plan also served to change the terms of the space race to favor the Americans. Russia's crude but effective throw-power would not get a manned ship to the Moon and back again because the task required sophisticated electronics, fine mechanical devices, and an ability to impose high standards of quality control—all things that were becoming American specialties.

On May 25, 1961, Kennedy addressed a joint session of Congress and asked for a national commitment to "landing a man on the moon and returning him safely to the Earth" within the decade. America's biggest, costliest, and most ambitious technological effort ever was under way, rivaled only by the building of the Panama Canal and the Manhattan Project's building of an atomic bomb in World War II.

Kennedy's proposal would not be cheap. Jim Webb, the new NASA administrator, was given a $10-billion cost estimate for the Moon project; he prudently decided to offer Congress a $20- to $40-billion estimate. (The program would ultimately cost about $20 billion.) Before coming to a decision, Kennedy had taken counsel with advisers who believed that the Moon project was feasible, largely because it could be "within the existing state-of-the-art" by expanding and extending the technology that already existed at that time.

However, since December 1957, when the first Vanguard had collapsed

John Glenn, the first American to orbit the Earth, practicing getting in and out of the Mercury space capsule. (NASA)

in flames before a television audience, the United States had tried to put twenty-five scientific satellites into Earth orbit; only ten were successful. The nation needed to improve its ability to reliably launch spacecraft in a hurry. When Kennedy announced his goal to Congress, the vehicle that would carry man to the Moon existed only as a theoretical concept tentatively named Apollo. The first Saturn rocket would not make its maiden flight for another six months.

NASA was up and running. Projects Mercury, Gemini, and Apollo had each begun and were all part of the grand plan. The Huntsville Rocket Team took responsibility for the Saturn launch vehicles, and the new Manned Spacecraft Center in Houston (created in mid-1962 but operating before that out of Langley, Virginia) was given responsibility for the payload—in this case the modules that would take the astronauts to the Moon's surface and back.

In February 1962, the American program moved forward when John Glenn orbited Earth three times and returned to a hero's welcome replete with a huge ticker-tape parade in New York City and command appearances at the United Nations, a joint session of Congress, and the White House. Glenn's orbit provided a hearty boost in national pride, making up for at least some of the earlier Soviet firsts.

By 1964 there were at least 250,000 people in the United States working directly or indirectly to fulfill the goal of a lunar landing. Triumph followed triumph, with manned spaceflight spectaculars sharing headlines with increasingly sophisticated unmanned spacecraft making the first quick visits to nearby planets. The American Mariner and Soviet Venera and Mars spacecraft went to Mars and Venus and gave the world its first close-up pictures of the surfaces of other planets. There were no canals on Mars, but there were huge craters created by meteors, the residue of volcanic eruptions, and snow and ice caps.

Through 1965 and 1966, the Gemini program provided the first U.S. space walks, space dockings, and pictures of the home planet from orbit. But after ten successful Gemini space-capsule launches in 1965 and 1966, an electrical fire on January 27, 1967, caused by a short-circuit, produced a spark that landed in an atmosphere of pure oxygen and produced a fire as intense as a blowtorch. The crew of Apollo 1—Virgil "Gus" Grissom, Ed White, and Roger B. Chaffee—preparing for what was to have been the first manned launch of the Apollo program, died in a matter of seconds. Grissom, a veteran of both the Mercury and Gemini programs, had been the second American to fly in space. He and John W. Young had flown the first manned Gemini mission, and NASA had slated Grissom to command the first flight of Apollo. White was also a veteran astronaut of note. During the flight of Gemini 4 with James A. McDivitt, White had become the first American to "walk" in space, aided by a handheld jet gun.

Shock gripped America after the accident. NASA administrator James Webb told the media at the time, "We've always known that something like this was going to happen sooner or later. . . . Who would have thought that the first tragedy would be on the ground?"

In the 1994 Turner Broadcasting System documentary series *Moon Shot*, an hour was dedicated to a thorough examination of the 1967 tragedy. *Moon Shot* surmised that the fatal fire on launchpad 34 was precipitated by U.S. competition with the USSR. The Soviet Union had been first to reach every space landmark up to that time, from Sputnik to Gagarin to Alexei Leonov's 1965 space walk. Rushing to catch up, says flight director Chris Kraft, "we'd gotten too much in a goddamn hurry." The other cause was the extraordinary complexity of the Apollo space capsule, which the narrator called "probably the most complex thing ever put together by humans. As much technology as a nuclear submarine crammed into a package the size of a minivan. The electrical system alone had 30 miles of wire."

The widespread assumption was that the race to the Moon was a surro-

The "Earthrise" that greeted the Apollo 8 astronauts as they came from behind the moon in lunar orbit. (NASA)

gate war with the Soviet Union—a contest on a new playing field. Before this race was over, both nations would push themselves as hard as they could, sometimes paying a devastating price. In October 1960 an unmanned Russian rocket designed to reach Mars exploded on the launchpad, killing 165 people, including Marshal Mitrofan Nedelin, an influential general. But this did not stop the Soviet Union any more than the loss of the three Apollo astronauts deterred the United States in 1967. If the space race was a type of warfare, the casualties were negligible compared to those resulting from armed conflict.*

The Apollo tragedy set the Moon mission back about eighteen months. During that time, NASA redesigned the ventilation system of the Apollo capsule. Russia meanwhile was delayed by its own tragedy, the death of Vladimir Komarov during the reentry of a test flight of the new Soyuz 1 spacecraft in April 1967. On October 11, 1968, Walter M. Schirra, Donn F. Eisele, and R. Walter Cunningham made the first manned flight of Apollo. That flight, Apollo 7, had been preceded by six unmanned test flights. But the Russians were still threatening with their own Moon-launch preparations.

*In his book *Little Wars,* H. G. Wells called for an elaborate war game that would be played in less than twenty-four hours. His contention was that "tin murder" could serve as a remedy for man's need for war. He wrote, "Let us put this prancing monarch and that silly scaremonger, all these excitable 'patriots,' and those adventurers and all the practitioners of *Weltpolitik* into one vast Temple of War, with cork carpets everywhere, and plenty of trees and little houses to knock down, and cities and fortresses, and unlimited soldiers, tons, cellarsful—and let them lead their own lives away from the rest of us."

An Apollo 11 astronaut's footprint on the Moon, photographed July 1969. (NASA)

Then came the bold decision that the Apollo 8 team would orbit the Moon. Apollo 8 launched on December 21, 1968, carrying Frank Borman, James A. Lovell, and William A. Anders. The crew was the first launched with NASA's new Saturn V launch vehicle—the apex of the Rocket Team's work. As it headed outward, the crew focused a portable television camera on Earth, and for the first time ever humanity saw its home from afar: a tiny, lovely, and fragile "blue marble" hanging in the blackness of space.

On Christmas Eve 1968, the world was stunned to hear an astronaut quoting from the Book of Genesis and see the view from our natural satellite.* Mythology scholar and teacher Joseph Campbell said of the Apollo image: "The earth is a heavenly body, most beautiful of all, and all poetry now is archaic that fails to match the wonder of this view. . . . Now there is a telling image: this earth, the one oasis in all space, an extraordinary kind of sacred grove, as it were, set apart for the rituals of life; and not simply one part or section of this earth, but the entire globe now a sanctuary, a set-apart Blessed Place."

To many, the image of the Earth as a spacecraft adrift in space rising above the Moon provoked a heightened concern with the environment. On

*Their orbit provided the first look at "Earthrise" as a bright, blue-marbled Earth rose gradually over the barren lunar surface set against the deep black of space. This famous "Earthrise" photo series was almost not taken. Frank Borman, the man in command, saw the beginning of the "Earthrise" and asked William Anders, the crew member in charge of photography, to get some pictures of it, but he refused, pointing out that it was not in his schedule of assignments. Borman grabbed the camera from Anders and took the remarkable series himself.

April 22, 1970, the first Earth Day was staged and declared as the beginning of the environmental movement. Not only did the Apollo 8 photos help inspire the event and the movement, but they also served as a reminder that there was a vast gulf between the image of a peaceful planet in space and the reality of life on Earth. The same month Apollo 8 orbited the Moon, the U.S. death toll in Vietnam passed the 30,000 mark.

In addition to affording a view of the dark side of the Moon, Apollo 8 showed the power of the Saturn V rocket. Before this, the highest altitude flown by a manned craft was 851 miles, but Apollo 8 went out to 240,000 miles.

Apollo 9 tested the lunar module by flying it while in Earth orbit. Then Apollo 10 orbited the Moon and tested its lander close to the Moon's surface. The testing was now over.

On July 16, 1969, with Neil Armstrong, Buzz Aldrin, and Michael Collins atop the massive Saturn V rocket, Apollo 11 blasted off from launchpad 39. On July 20 at 4:17 P.M., Washington time, Armstrong radioed back: "Houston . . . we, uh . . . Tranquility Base here. The Eagle has landed."

After a six-hour sleep, just about four minutes before 11:00 P.M., he became the first human to step on the Moon. "That's one small step for man, one giant leap for mankind," he said, marking the event for posterity. The rest of the mission was flawless, and the world celebrated the wonder of it all.

"Of all humankind's achievements in the twentieth century—and all our gargantuan peccadillos as well, for that matter—the one event that will dominate the history books a half a millennium from now will be our escape from our earthly environment and landing on the moon," wrote Walter Cronkite in his 1996 book, *A Reporter's Life*. Sputnik and the intensity of the American reaction to it, more than any other single event, marked the start of the space race; the Apollo 11 mission to the Moon ended it.

CHAPTER TEN

Sputnik's Legacy

If I could get hold of that thing, I would kiss it on both cheeks.

—Major General John B. Medaris, discussing Sputnik several years after the fact

Prior to the launching of Sputnik I, the United States had enjoyed a period of economic growth and national pride that began at the end of World War II. The nation was buoyed by its resolve, courage, and confidence in its scientists', engineers', and technicians' ability to create stunning technological advances on short notice—radar, sonar, a very effective pesticide (DDT), large-scale production of synthetic rubber, the atomic bomb, and nuclear power. But Sputnik changed that.

Gabriel Heatter, an influential news commentator for the Mutual Broadcasting System, delivered a radio editorial titled "Thank You, Mr. Sputnik" in January 1958, hours after Sputnik I had fallen from orbit. After reporting its demise, he addressed an object that no longer existed: "You will never know how big a noise you made. You gave us a shock which hit many people as hard as Pearl Harbor. You hit our pride a frightful blow. You suddenly made us realize that we are not the best in everything. You reminded us of an old-fashioned American word, humility. You woke us up out of a long sleep. You made us realize a nation can talk too much, too long, too hard about money. A nation, like a man, can grow soft and complacent. It can fall behind when it thinks it is Number One in everything. Comrade Sputnik, you taught us more about the Russians in one hour than we had learned in forty years."

Within weeks America had begun to use Sputnik to reinvent itself. Using the metaphor of the hour, science-fiction editor John Campbell observed, "There is nothing like a good, hard kick in the pants to wake up somebody who's going to sleep on the job." Many sectors wanted to direct society along new lines, and it seemed that everyone had an idea on how this could be accomplished. Legislators had all sorts of ideas: Representative Earl Wilson (R-Indiana) said that Sputnik proved that the nation needed a "West Point for the Sciences" to train scientists and engineers. Representative Kenneth B. Keating (R-N.Y.) called for a "Manhattan Project of international dimensions to co-ordinate and bring to perfection the satellite missile projects of all the North Atlantic Treaty Organization nations."

Others thought that Sputnik was a call for giving new life to old institutions. For instance, a special 1958 issue of *Scouting* magazine, aimed at the leadership of the Boy Scouts of America, ran a series of three Sputnik-related articles. The first began: "The best thing that has happened to the United States in our generation is Russia's Sputnik. It is proving the most effective therapy conceivable for our indifference and negligence, our concentration on comforts and convenience, our devotion to all but our patriotic obligations in this best of all lands." The author of this article, Louis B. Seltzer, editor of the *Cleveland Press,* weighed in with the argument that the nation had been spending billions of dollars on things ranging from highways to "expensive professional sports stadiums," money which would be better spent on "an organization like the Boy Scouts and similar organizations for the girls." If such a redistribution of wealth were to occur, "the future of America would be secure against anything that could possibly challenge us in the future." There was a redistribution of wealth, but the money did not go to scouting.

Sputnik's impact on the technological and scientific community in the United States was monumental. It led to the creation of the Advanced Research Projects Agency (ARPA) to ensure that the United States would never again be left behind in the area of new technology and the beginning of flush times for the National Science Foundation. The foundation's annual budget rocketed from less than $50 million for 1958 to $136 million the next year, and doubled again by 1962. It is now over $1.6 billion.

Just as Sputnik spurred the space race, it also spurred a race to develop and produce high-tech weapons. As former ambassador to the Soviet Union George Kennan put it in his memoirs: "It caused Western alarmists . . . to demand the immediate subordination of all other national interests to the launching of immensely expensive crash programs to outdo the Russians in this competition. It gave effective arguments to the various enthusiasts for nuclear armament in the American military-industrial complex. That the dangerousness and expensiveness of this competition should be raised to a new and higher order just at the time when the prospects for negotiation in this field were being worsened by the introduction of nuclear weapons into the armed forces of the Continental NATO powers was a development that brought alarm and dismay to many people besides myself."

The arms and space races, however, were just part of the reaction. The first big benefactor outside the federal government itself was education.

Upgrading the 3 *R*s

During the early 1950s a series of bills in support of public education had been introduced in Congress, but not a single one reached the president's desk for signature. Ever since the New Deal, federal aid to education had been fought successfully by a powerful coalition composed of southerners, who were afraid it would be used to force desegregation; fundamentalists, who feared the encroachment of Darwinism; Roman Catholics, who worried that no benefit would accrue to parochial education; and conservative Republicans, who did not want the federal government intruding into a new realm. But after Sputnik, as Peter B. Dow wrote in his 1991 book *Schoolhouse Politics: Lessons from the Sputnik Era,* "the public demanded to know why our space scientists had failed to keep pace with the Soviets, and many critics were quick to place the blame on inferior schooling. In the mid-1950s the popular press teemed with articles extolling Soviet educational practices and questioning our own."*

The press was the first to point to education as the culprit. Harry Schwartz, a reporter who covered Russia for the *New York Times,* led off on October 14 with an article stating that educational drive and the prestige accorded science students underlay the Soviet triumph. *Life* magazine then carried a laudatory article pointing out that the Soviet system operated on a basis "which draws or forces all human knowledge into service of the state."

A chord had been struck, and before the year 1957 was over the Gallup pollsters announced that 70 percent of those surveyed agreed with the as-

*In March 1956 *Life* published a laudatory article on Soviet science education, praising the "golden youth of communism," and *Scholastic* magazine called the competition with the Soviet Union a "classroom cold war."

sertion that American high school students must now work harder to compete with the Russians.

Throughout 1958, Congress held numerous hearings on questions of science and education. In a hearing before a subcommittee of the House Committee on Education and Labor, representatives questioned their star witness Wernher von Braun on his recommendations for improving America's scientific workforce. Von Braun emphasized the need for further development of human resources in the sciences: "Modern defense programs are the most complex and costly, I suppose, in the history of man. Their development involves all the physical sciences, the most advanced technology, abstruse mathematics and new levels of industrial engineering and production." He called for a new kind of soldier, one who might one day be memorialized as the man with the slide rule. "It is vital to the national interest that we increase the output of scientific and technical personnel." Science, which had less than a decade before reached a low in public perception, now had taken center stage. Von Braun's testimony before the committee was heartily applauded—something extremely rare in the congressional hearing room.

The reason given for America's educational deficits varied according to who was making the assessment. A National Education Association spokesman noted that "any nation that pays its teachers an annual average of $4,200 cannot expect to be first in putting an earth satellite into space." The director of the American Institute of Physics said that the American way of life was "doomed to rapid extinction" unless the nation's youth could be taught to appreciate the importance of science. As NASA archivist Lee Saegesser noted twenty-five years after the fact: "Frankly, it was a field day for certain interest groups. One could get quoted in the papers or invited to testify in front of a Congressional committee just by coming up with some statistic which showed how bad our schools were compared to those in Russia."

In a hearing before the Senate Committee on Labor Relations and Public Welfare, the chairman of the National Science Board, Detlev Bronk, outlined a solution to the science problem that would provide more federal support for research and education in the sciences, increase the study of the hard sciences in elementary and secondary schools, and change the educational system's emphasis from rote memorization to creative thinking. From that moment on, the rallying cry was banish rote learning.

The only member of the committee who opposed a federal program to implement Bronk's suggestions was Senator Strom Thurmond of South

Carolina. He cited fiscal and constitutional grounds for his disapproval. The rest of the committee, including powerful rising political stars such as Senators John F. Kennedy of Massachusetts and Barry Goldwater of Arizona, supported the program even if it meant more taxes.*

The biggest result of the House and Senate hearings was the passage in 1958 of the National Defense Education Act (NDEA) "to meet critical national needs," a bipartisan piece of legislation if ever there was one. NDEA appropriated $47.5 million in student loans for 1958, with expenditures on loans budgeted to exceed $100 million by 1962. Also, over a four-year period, nearly $300 million went to fund the purchase of scientific equipment and the establishment of National Defense Fellowships for graduate students.

On September 2, 1958, the president signed into law the National Defense Education Act, "an emergency undertaking to be terminated after four years . . . to bring American education to levels consistent with the needs of our society." This was the first time since 1917 that serious attention was being paid to school reform. NDEA poured billions of dollars into the educational system over the next decade to pay for language labs, the "new math," and the broad curriculum overhaul of the late 1950s and early 1960s. The funds were for high schools as well colleges and universities. And the aid for science and math also spurred increased support for liberal arts. English majors could get loans provided by the National Defense Education Act, part of which would be "forgiven" if they went into teaching. The federal government had committed itself to shoring up the public schools and giving a major boost to colleges and universities.

Between 1958 and 1968, NDEA also provided loan money for more than 1.5 million individual college students—fellowships directly responsible for producing 15,000 Ph.D.s a year. NDEA allocated approximately $1 billion to support research and education in the sciences over four years; federal support for science-related research and education increased between 21 and 33 percent per year through 1964, representing a tripling of science research and education expenditures over five years. States were given money to strengthen schools on a fifty-fifty matching basis, thousands of teachers

*Another element that came into play at this time was competition with the Russians—that education was one more playing field on which Americans had to come from behind and win. On June 13, Lawrence G. Derthick, U.S. commissioner of education, reported to the National Press Club on his fact-finding tour of the USSR. He said that education in Russia is a "kind of grand passion" and that the Soviet Union has a "burning desire to surpass the United States in education, in productivity, in standards of living, in world trade—and in athletics."

were sent to NDEA-sponsored summer schools, and the National Science Foundation sponsored no fewer than fifty-three curriculum development projects. By the time of the lunar landing in 1969, NDEA alone had pumped $3 billion into American education.

Patrick Lackey, a writer for the *Virginian-Pilot* (Norfolk), was one of the beneficiaries. He told of his NDEA blessing in a 1997 article for his newspaper: "I, a senior at Belleville High School, was handed a science scholarship. . . . I had taken a couple of science courses, but I'd never mentioned to anybody that I had any interest in science. In fact, I had no idea what chemists did, other than mix liquids in test tubes and observe as they changed color. None of that mattered. The nation was in such a panic that science scholarships were being handed out like candy. In the rush to turn American youths into scientists, mistakes were made. I was one of them. After a semester and a half as a chemist, spending entire afternoons in a lab, I switched my major to music. But darn it, America was first to the moon, eight years later."* Among Americans who were in the educational system in the NDEA years, there are countless other stories. Duffy White, now a professor of Russian literature at Wesleyan University in Middletown, Connecticut, is one of a sizable cohort of NDEA beneficiaries who can say without the slightest hesitation: "Sputnik changed my life."

Women comprised one subset of NDEA beneficiaries. Although feminism still had no name in 1957, its roots were in place. It was bolstered by Western revelation that the Russians were using the female half of the population to effect scientific and technical advances—a point underscored on June 16, 1963, when Valentina Tereshkova became the first woman in space. The other catalytic force was the NDEA, which put out a welcome mat for women to enter scientific and technical fields.†

*One of the beneficiaries of NDEA was Theodore Kaczynski, the Unabomber, who moved to the University of California, Berkeley, in 1967 as an assistant professor of mathematics. This was a time when government funding for science and math programs was peaking in response to the launch, a decade earlier, of Sputnik. Berkeley's math department had seventy faculty spots in 1967, making it easier for Ph.D.s to land a professorial spot.

†Feminist Betty Friedan wrote in the January 3, 1994, *Newsweek* that in 1957 she had created a questionnaire for the alumnae reunion of the Smith College class of 1942 which she hoped to sell as a magazine article. "I wanted to use the results to prove that education was good for women—that it hadn't made them frustrated as wives, mothers and homemakers." She found that many women were restive and unfulfilled, and nobody wanted to publish her article. Friedan kept interviewing women with college educations, and her findings ultimately appeared in her book *The Feminine Mystique*.

Beyond the NDEA, there were major changes in the way courses were taught. One of the most effective was the Biological Sciences Curriculum Study; it revolutionized the teaching of biology, as well as the textbooks. Mike Bowler, longtime education editor of the *Baltimore Sun,* described it: "Students no longer memorized biological facts systematically. They considered the process of life's development and they got a heavy dose of evolutionary theory which BSCS called the 'warp and woof' of modern biology."

Charles Darwin and his 1859 theory of evolution had been successfully kept out of many classrooms until late 1957, when Sputnik panicked the scientific establishment and the theory of evolution found a place in high school biology textbooks. An article in the *St. Paul Pioneer Press* (Minnesota) of September 18, 1999, commented, "Though some protested, they couldn't for long. The late 1950s and early 1960s was a time of anti-communist rhetoric: To be against science was to be against the technological race against the Russians."

Schools placed new emphasis on the process of inquiry, independent thinking, and the challenging of long-held assumptions. Laboratory science was stressed, urging a hands-on learning approach. The emphasis moved from teaching facts to fundamental principles. America's children could no longer be educated traditionally. Latin and Greek were suddenly out of favor; language labs and fluency in modern languages were the order of the day.

More attention also was paid to homework. Journalist, literary agent, and Hollywood producer David Obst recalled October 1957 in his autobiography: "I was in the sixth grade and I thought school was just fine. I liked my teachers, the homework wasn't bad. Then Sputnik. . . . Suddenly my teacher brought this newspaper into class. She read us the following: 'Russian pupils go to school six hours a day, six days a week, attending classes 213 days a year compared to a mere 180 in the United States, and have FOUR HOURS of homework each day.' [My] classmates and I became staunchly anticommunist." Obst and his classmates were given to believe that the only way that America was to prevail was by loading on the homework. "My friends and I hoped this would all go away, but just before Thanksgiving the goddamn Russians shot off Sputnik II and put a little dog or monkey or something into the cosmos. Now we were really in for it," Obst added.

Typically, schools and school systems would impose their own programs on top of what was being mandated from Washington and the statehouse. Public schools in Pleasantville, New York, had a program known as "Operation Upgrade," which was described in a 1959 paper by high school senior Douglas Evelyn as "Pleasantville's answer to the national problem of keeping the education of today's younger generation in step with that of Russia." Evelyn, now an administrator at the Smithsonian in Washington, cataloged all the changes that had been wrought in his school in the sixteen months since Sputnik first orbited. There were honors courses in chemistry and other sciences; math courses at an earlier age, giving exceptional students the chance to complete geometry, trigonometry, and other classes in a half rather than a full year; enriched and more intense courses in German and French. More honors courses were added as the system rallied to "meet the Soviet challenge." Students were now grouped by intelligence, and, to Evelyn's dismay, vocational courses were given short shrift.*

In the wake of Sputnik, many colleges and universities became rich—especially those that performed research on a large scale in such fields as lunar and planetary studies, plasma physics, and astronomy. The way in which university research was conducted changed with the federal largesse. Stressing the need for this change in 1958, Arthur R. von Hippel of MIT wrote that the shock of Sputnik "spells the end of the specialization of the past, where scientists and engineers of the various professions were walled up from each other in air-tight subdivisions of schools and departments. The time of synthesis of specialized knowledge has arrived, in which we begin to think about materials and their applications in unified vision." The new model for serious accomplishment was not the university department but the "think tank"—a freewheeling, interdisciplinary band of problem solvers such as the group working for the Air Force's RAND Corporation.

Attitudes toward intellectuals changed virtually overnight. "All at once the sleazy downgrading of people who lived by books and words as 'eggheads' sounded archaic," observed Paul A. Carter in *Another Part of the Fifties.* "Suddenly the national distaste for intellect appeared to be not just a disgrace but a hazard to survival," said historian Richard Hofstadter.

Clark Kerr, then president of the University of California, remarked in 1964 that, ironically, America's universities, "which pride themselves on their autonomy . . . which identify themselves either as 'private' or as 'state' should have found their greatest stimulus in federal initiative—that institu-

*The deemphasis of shop and other practical classes after Sputnik had its own irony. At the turn of the twentieth century, there had been growing concerns about America's ability to keep up with competitors. The worries helped give birth to the Smith-Hughes Act of 1917, which called for a vocational education curriculum in the nation's public schools.

tions which had their historical origins in the training of 'gentlemen' should have committed themselves so fully to the service of brute technology." The effects of this change are still in place, as universities continue to rely on federal research money, which in turn follows the policy of the agency handing out the grants.*

Brave New Thinkers

The Baby Boomers (born between 1943 and 1960) were the ones who went on to become the inquisitive students of Sputnik-era and post-Sputnik schools. Raised on Doctors Spock and Seuss, they were taught to think independently, question authority, and—above all—do more homework.†

Suddenly, students were being encouraged to read outside the realm of the assigned text, to look at alternative theories to venture closer to the dangerous far edge of conventional thinking. Publisher Cass Canfield recalled in his memoir that Sputnik ushered in the day when trade books supplemented basic texts in college curricula. Even at the least progressive schools, students were being encouraged to go beyond the distilled version and right to the original, whether it be Marx, Freud, or the Federalist Papers. Before long, the hot books on campus tended to be works like Herbert Marcuse's *One-Dimensional Man,* which argued that mass materialism was stifling diversity in American culture, Theodore Roszak's self-

*From the start there were those who saw something negative at work in the Sputnik reaction and the new emphasis on science accentuating a long-standing academic/intellectual feud between science and the humanities. C. P. Snow, a scientist and novelist, straddled the "gulf of mutual incomprehension—sometimes hostility and dislike" that divided scientists, to whom the entire literature of the traditional culture seemed either regressive or not relevant, from "the literary intellectuals who . . . while no one was looking took to referring to themselves as 'intellectuals' as though there were no others." The chief executive officer of the American Association for the Advancement of Science, Dael Wolfle, reported in 1959 that "it was not uncommon to hear someone say with smugness and even a touch of pride, 'I don't know a thing about science.' "

†Published in the spring of 1957, the 225-word primer *The Cat in the Hat,* by Dr. Seuss (Theodor Seuss Geisel) suddenly became a best-seller in the wake of Sputnik. Reviewing a book about Seuss in the *New York Times,* Ann Hulbert noted that "the peculiarly named 'doctor' and his upstart reading primer, an antidote to dull Dick and Jane, were at the forefront of the crusade to make American children cognitively, and creatively, competitive."

descriptive *Making of a Counterculture,* and Eldridge Cleaver's angry *Soul on Ice.*

These educational changes were dramatic and obvious, especially to those who grew up in a well-ordered universe that embraced Latin, Dick and Jane primers, classical literature, and rote learning—and then swung into a world of freewheeling creativity where the new sacrament was "drawing one's own conclusions." People who grew up with a cold impersonal subject called biology, with its emphasis on one-celled animals and memorizing names of the bones of the body, watched their younger brothers and sisters study something they called "life science," which viewed life as something greater than the sum of its parts. Pre-Sputnik students were told about the classic experiments of science, but post-Sputnik kids learned firsthand through their own experiments, the best of which were displayed at science fairs.

A new trend toward independent thinking spurred rebellion and activism. The civil rights movement, an unpopular war in Vietnam, and the empowerment of a distinct youth culture fostered by rock 'n' roll led to a generation populated in part by flower children, antiwar protestors, draft resisters, dropouts, Jesus freaks, and hippies. Ten years after Sputnik, students questioned the morality of the war in Southeast Asia. They wondered why alcohol had the blessings of the establishment, but not marijuana. They held their universities accountable for doing all sorts of unsavory weapons work for the Pentagon while turning a blind eye to the problems of starvation, disease, and poverty.

Randall Balmer, associate professor of religion at Barnard College, has theorized that Sputnik and the 1960s had a causal link to today's worldwide return to spirituality. In a November 13, 1993, Religious News Service article, he wrote: "My own sense is that the surge of interest in religion has a lot to do with the failure of science and technology, which, you'll recall, was supposed to be our savior. After the Soviets launched Sputnik in 1957, the United States began a heroic effort to teach science in the public schools. 'Technology was supposed to solve all of our needs.' But something, somewhere went wrong. By 1969, when the United States placed its man on the moon, Americans were becoming disillusioned with technology. Young people, the members of the counterculture, began to sense that technology was dehumanizing, and they called into question our uncritical embrace of 'progress.' . . . Soon others began to have doubts. Technology, our supposed savior, had brought us smog and congestion, Three Mile Island, and Chernobyl nuclear disasters. Technology had given us microwave ovens and electric can openers, but it hadn't taught us how to live, how to imbue our lives with meaning."

Sputnik inspired the generation that put a man on the Moon and, to some extent, the next generation. "I was born ten days before the Sputnik went into orbit," said Jim Waterman on National Public Radio's *Sounds Like Science* for March 6, 1999. "I watched with wonder as the Gemini missions and Apollo moon missions took place. And so I can relate to the excitement and urgency of the American public as these breakthroughs happened. As a kid, I found myself dreaming of being an astronaut. I was always experimenting with electronics, motors and blowing things up. As I grew up, I realized these were the kind of things that engineers get to do, and I decided to become one. So I've been privileged to lead construction of the International Space Station."

Homer Hickam Jr. first saw Sputnik's dim pinprick of light moving across the night sky over Coalwood, West Virginia, when he was fourteen years old. The sight changed his life. He became an engineer at NASA; he worked on the Hubble Space Telescope, among other projects, and wrote the story of how he got there. That story, *Rocket Boys,* was published in 1998 and then turned into the movie *October Sky.* In an NPR *Weekend Edition* segment in 1999, he said: "When we walked out in the backyard and saw Sputnik fly over, to me it was the most amazing thing because my whole life had been really separated into two parts. That is, everything that happened in Coalwood and everything that happened in the outside world. And all of a sudden, there was the outside world over Coalwood, West Virginia, USA, and to me, that was the most amazing thing."

Film critic Roger Ebert, reviewing *October Sky,* shared his own recollection: "I remember the shock that ran through America when the Russians launched Sputnik on Oct. 4, 1957. Like the residents of Coalwood, West Virginia, in the movie, I joined the neighbors out on the lawn, peering into the sky with binoculars at a speck of moving light that was fairly easy to see. Unlike Homer Hickam, I didn't go on to become a NASA scientist or train astronauts. But I did read Willy Ley's *Rockets, Missiles and Space Travel* and Arthur Clarke's *The Making of a Moon.* I got their autographs, too, just as Homer sends away for a signed photo of Wernher von Braun. That first shabby piece of orbiting hardware now seems like a toy compared to the space station, the shuttle and the missions to the moon and beyond. But it

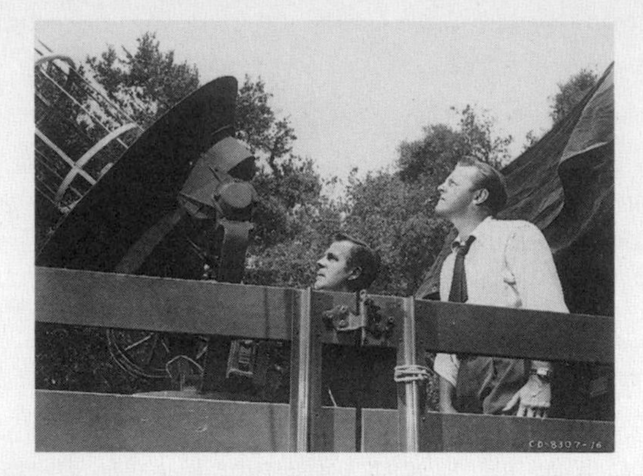

Scene from the 1956 movie *Earth vs. the Flying Saucer,* seen by an impressionable Stephen King on the night Sputnik went into orbit. In the movie an alien scouting party arrives at a U.S. Army base in search of help for their dying planet, but the military fires on their fleet of flying saucers and the aliens are forced to retaliate. (Museum of Modern Art Film Still Archives)

had an enormous impact. For the first time in history, man had built something that went up, but did not come down—not for a long time, anyway."

Sputnik also inspired individuals to embark on careers in international relations. One notable example is Strobe Talbott, deputy secretary of state and international negotiator for the Clinton administration, who told interviewers that his interest in Russia and international affairs dated back to the launch of Sputnik when he was eleven.

Budding writers seem to have been particularly affected by Sputnik. Stephen King, spinner of graphic tales of fear, vividly recalled that "for me, the terror—the real terror, as opposed to whatever demons and bogeys which might have been living in my own mind—began on an afternoon in October of 1957." King had just turned ten and was in a movie theater attending a matinee of *Earth vs. the Flying Saucers* in Stratford, Connecticut, when the theater manager shut off the projector to make a somber, eerily suitable, real-life announcement: Sputnik had been launched. The line between film horror and real horror blurred for King that day: "I am certainly not trying to tell you that the Russians traumatized me into an interest in horror fiction, but am simply pointing out that instant when I began to sense a useful connection between the world of fantasy and that of what *My Weekly Reader* used to call Current Events."

In his autobiography, *In Joy Still Felt,* Isaac Asimov reported his reaction to the news: "I berated myself for spending too much time on science fiction when I had the talent to be a great science writer. . . . From that time on, it was science that chiefly interested me, and though I continued to write science fiction now and then, it was only now and then. Never again, after the fall of 1957, was science fiction to form the main portion of my output." The late writer of more than 100 books added, "Sputnik also served to increase the importance of any known public speaker who could talk on science and, particularly, on space."

Anniversaries

While the shock of Sputnik I was immediate, it took many years for its effects to be fully known. Anniversaries have been occasions for thoughtful reevaluation of the satellite. In October 1977—exactly twenty years after the fact—with an American flag standing on the Moon, the *Washington Post* noted in an editorial that Sputnik had created a generation of engineers and scientists unmatched by any other in "size and skill." The *Pittsburgh Press* applauded Sputnik's impact on the microminiaturization of electronics, which made possible "hand-held computers that can perform functions which 20 years ago would have required a computer the size of a house."

The twenty-fifth anniversary of Sputnik brought forth a massive outpouring of exhibits, conferences, reminiscences, and nostalgia for the world of 1957, complete with images of backyard bomb shelters, Ike on the golf links, and Elvis Presley in convict's stripes wailing "Jailhouse Rock." The most widely quoted estimate was that the two superpowers had sunk $300 billion into space and that the Earth had gotten back a fair amount in return. Scientists had learned more about the universe in the last quarter century than in all the time before 1957.

But the twenty-fifth anniversary also stirred a renewed feeling of dread because of the arms race. "With laser 'death rays' on drawing boards and 'killer' satellites already in orbit, a real-life *Star Wars* has commenced between the U.S. and the Soviet Union," wrote Jim Detjen in his 1982 article "Sputnik Plus 25: Death Rays and Lasers" for the *Philadelphia Inquirer.*

American newspapers were full of the accounts of Soviet killer satellites, particle beam weapons, and the need for American satellites that could catch up with them. Harold Brown, secretary of defense in Jimmy Carter's administration, warned that the United States had to do everything it could to avoid turning space into a war zone. A party at the Soviet Embassy celebrating the twenty-fifth anniversary was boycotted by NASA invitees, according to the *New York Times.* Soviet astronauts were flying overhead on a daily basis in what *Aviation Week & Space Technology* termed "at least the embryo of a permanent manned space station."

U.S. relations with the Russians were as tense as they had been in 1957 and nowhere near as friendly as they had been in 1975 with the joint Soviet-U.S. docking mission. "Ironies abound in what Sputnik wrought," wrote Thomas O'Toole in the *Washington Post* on October 4, 1982. "Kennedy never saw the flight of anything to which he gave the initial impetus, and President Johnson, who did more for the Apollo manned program than any president, suffered through the fatal Apollo fire. President Nixon inherited and basked in all of the Apollo triumphs only to answer for Watergate."

The mood was quite different for the thirtieth anniversary in 1987. President Ronald Reagan's proposed "Star Wars" plan (officially "Strategic Defense Initiative," or "SDI") to intercept Soviet missiles acted as a bluff in the nuclear poker game, and the Russians were backing off their threat. NASA's space shuttle fleet remained grounded in the wake of the January 1986 Challenger accident. But when the Soviet Union's Space Research Institute sponsored a forum on the future of space, NASA sent a delegation.

"Thirty years after the Sputnik launch, both the Soviets and the United States remain in the forefront of the exploration of space," wrote Mark Carreau in the *Houston Chronicle.* "U.S. leadership, marked by the spectacular feats of the Apollo and the early successes of the space shuttle, has been supplanted by the more methodical pace of the Soviet program, now symbolized by its Mir space station—a permanent home for orbiting cosmonauts."

At this point the differences between the two programs were underscored. Although the Soviet Union was first to go into space, it concentrated on simple technology and the ability to launch scores of rockets each year. America's Apollo and space shuttle programs, meanwhile, relied on only a few launches but involved a more dazzling technology, which U.S. officials did not want to share with anyone, especially not the Soviets. When the space shuttle was first launched in 1982, it outshone anything the Soviets had developed. But since the Challenger disaster in January 1986, the Soviets had boasted about the benefits of their lower-tech strategy. The Soviets were able, for instance, to launch two missions to investigate Halley's Comet in 1985, joining spacecraft from Japan and Europe. The biggest Soviet plum, launched shortly after the Challenger explosion, was the modular space station Mir, which gave Russia a long-lived presence in space.* Nevertheless, the cash-poor National Aeronautics and Space Administration was still able to plan and launch highly successful space missions to Venus, Mars, and the outer planets. The two Voyager missions sent back thousands of the most beautiful and informative pictures ever received of Jupiter, Saturn, Uranus, Neptune, and their satellites. The space shuttle was used to place the Hubble Space Telescope into orbit. It brought breathtaking views of the galaxies and nebulae scattered in clusters across an expanse so large that light would take 20 billion years to travel through it.

By the time of the thirty-fifth anniversary, the Soviet Union was no more. The dissolution of the USSR was poignantly illustrated by the following Reuters dispatch:

> MOSCOW—The lost-in-time Soviet cosmonaut hurtled back to Earth March 25, 1992 after circling the planet more than 5,000 times and spending 7,512 hours in space for his country that no longer exists. Cosmonaut 3rd class Sergei Krikalev earned a minute on Russia's television news. He blasted off from one country—the Union of Soviet Socialist Republics—in May 1991. He landed in the sovereign nation of Kazakhstan—one of the 15 countries that emerged from the U.S.S.R. When he left last May, a ruble was still a ruble. Today it's worth about a penny, and Mr. Krikalev's 500-ruble-a-month salary is worth about $5.

The story of Sputnik seemed complete on March 13, 1995, when Norman Thagard, a veteran of four shuttle flights, became the first American to go into space on a Russian rocket. His launch was part of the Mir program, joining an illustrious group of space travelers to have departed Earth from a launchpad at the once-secret Baikonur Cosmodrome in the steppes of central Asia. Yuri Gagarin, the first man in space; Valentina Tereshkova, the first woman in space; Alexei Leonov, the first spacewalker; Laika, the first dog in space—all left from this same spot, which Russians reverently call Gagarin's launchpad. The first of more than 300 launches to have lifted from this pad was Sputnik in 1957.

At the fortieth anniversary, attention was paid to what had happened to

*Mir returned to Earth on March 23, 2001, after more than fifteen years in orbit.

Russia's Mir space station, as photographed from the approaching U.S. space shuttle Atlantis on June 29, 1995. (NASA)

the planet between 1957 and 1997, not the least of which was that approximately 4,800 satellites and space probes had been successfully launched.* Russia was now selling satellite services to other nations. Without Russian rockets for hire, there would not be enough launch capacity for less developed nations, many of them lacking even basic telephone services, to leap forward with satellite communications. Significantly, the fortieth anniversary year was the first year in which there were more commercial than military satellites in orbit. Those commercial satellites provided a platform supporting telephone, television, the Internet, weather forecasting, and, by extension, banking and finance.

Sputnik was getting its full due at forty. NASA's own *HQ Bulletin* termed it "a symbolically significant achievement which affects our lives today." An article in the *Wall Street Journal* by Pulitzer Prize–winning historian Walter A. McDougall termed it the "greatest watershed in the history of the Cold War." Others marked the changes that had occurred in terms of stunning contrast. For instance, the average junior-high-school math class

*The *Chicago Tribune* noted that if one included debris—such as tools dropped by astronauts—there were by this time 10,000 man-made objects in space. One of the fortieth anniversary's disquieting notes was contained in physicist Bob Park's October 3 newsletter: "SPUTNIK: 40TH ANNIVERSARY CELEBRATED BY ZAPPING A SATELLITE? The space age began on Oct 4, 1957 with the launch of Sputnik. Well, let's not get all mushy about it, the Pentagon wants to mark the occasion by blasting an Air Force satellite out of the sky with an infrared chemical laser left over from SDI. The Soviet Union is history, nuclear arsenals are being dismantled, and SDI is dead, but the laser lobby lives on."

of twenty students in 1997 used personal computers with more computing power than the Pentagon did at the launch of Sputnik in 1957.

While almost everything that makes it into print or is broadcast assumes that Sputnik's effects were all good, there are some thoughtful people who do not agree. Beginning in the late 1980s, a number of people strongly suggested that U.S. reactions to Sputnik had some negative aspects. One argument was that it had created a monster of mock science, which dictated that everything—even common sense—be tagged with a coefficient or correlation or some other numerical value that included a decimal point. Critics in various fields held that Americans' intuitive feel for the world was being lost in a new world where every problem seemed to need a "scientific approach."

By the mid-1990s some argued that the effect of Sputnik had been to take people away from the pragmatic and the utilitarian, bringing about a revolution in training that greatly altered the teaching and practice of engineering. William S. Butcher, a senior engineering adviser at the National Science Foundation, which had encouraged the revisions decades before, recalled that "before [Sputnik], engineering students were taught how to make a road; you dealt with people who had experience, who had built roads or bridges. . . . Students were shown things and shown how to do it." But after Sputnik, the whole discipline became more theoretical and abstract, more concerned with the examination of physical principles and less with teaching students how things are actually built.

All of this came at a time when Americans were told that they were witnessing the deterioration of the national infrastructure, the roads, bridges, pipelines, wires, and communications networks. To be sure, the world could become completely wireless at some point in the future, but people could not go sewerless. The critics charged that the shift away from the basics was underscored by a number of very visible engineering failures of the previous decade or so, including defects in the Hubble Space Telescope, the USS *Stark*'s flawed defense system against an Iraqi missile, the collapse of the Hartford, Connecticut, Civic Center (under a load of snow), and the leak at the Three Mile Island nuclear plant. Critics called for new emphasis on "pragmatic training" and the reintroduction of design in the early stages of the curriculum.

Eugene S. Ferguson, author of *Engineering and the Mind's Eye,* and an emeritus history professor at the University of Delaware, not only faulted the turn away from the pragmatic side of engineering and especially the deemphasis on design for such problems; he recently joined those who warn against what they regard as an overreliance on computer software in the design process. These trends have contributed to spectacular engineering failures.

Even the money pumped into science was seen as having a dark side. One of the most vocal critics of the post-Sputnik spending revolution was Joseph Martino, who held that dramatically increased funds undermined the culture of scientists as idealistic intellectuals pursuing knowledge. He wrote in *Science Funding: Politics and Porkbarrel* that the effect of increased science funding was to destroy science as it had existed before and to turn scientists into wards of the state, chasing grants.*

Sputnik, it has been argued by a band of critics, was actually a detriment to the study of math and science in the United States. So-called new math proved marginally relevant and didn't survive long in the classroom. When the Soviets launched the first satellite in 1957, there was a rush to sign up students for science courses, but schools couldn't take all who wanted to participate so they made courses excessively tough. James Cook, director of an initiative to improve science in the schools called "A World in Motion," was quoted in November 1, 1985, by the Associated Press: "It was a turn-off when a kid's successful parent couldn't figure out a homework science problem. . . . It didn't take a generation of kids long to say, 'Who needs this?' "

Jon Miller, political science professor at Northern Illinois University and vice president of the Chicago Academy of Sciences, told an NPR *Talk of the Nation* audience on October 3, 1997, that within the science education community, there were already "several initiatives that were rewriting physics, rewriting chemistry and rewriting biology to try to get students to think more in an inquiry mode; to understand the rationale behind science; to understand how to think about it in a way that a scientist might think about it, as opposed to simply memorization." Miller said that the result of these reforms was some remarkable work. But when asked what the long-term results had been, he replied: "Well, what it did, inadvertently, was it truncated or polarized our educational system. When you began to ask high school students to think about more abstract ways of looking at science,

*Garrett Moritz, in his article "The Changing Role of Science During the Cold War," argued, "If turning scientists into entrepreneurs first and researchers second has the effect of improving the rate of scientific progress and improving the material world, Martino's criticism is valid only from a nostalgic perspective."

some of the students who have come through the elementary experience were ready to do that and some were not. And inevitably, as those curricula came into broad use, some high school students began to fail. And that is, of course, in terms of the stability of a high school, a very bad thing."

Space-Age Products and Services

In September 1956, before Sputnik was launched and NASA came into existence, a California Buick dealer wrote to the Air Force attempting to buy exclusive advertising space on all space vehicles, which he understood would start flying in the next year or so. He was turned down, and the historic exchange of correspondence now rests in a file in the NASA Archives in Washington, D.C. However, the Buick dealer's vision was prophetic in the sense that capitalism ultimately would benefit from space exploration.

One of the great myths of the Space Age is that it brought the world little in return. In other words, billions were spent, and all that came from it was "Teflon, Tang and Velcro."* In reality, space led to numerous products and processes, many of which have shaped everyday life immeasurably. The space race forced America to push itself into new realms that it otherwise might have explored decades later.

Engineering and scientific advances linked to space include fuel cell technology, which uses oxygen and hydrogen to produce potable water and electricity. Tough, antiflammable fabrics, such as Kevlar, were developed and used for space suits and backpacks. Space-age biomedical instrumenta-

*Teflon was invented by accident in 1938 by Roy J. Plunkett, a chemist in the DuPont labs who saw that this residue of refrigeration gases had unusual properties. First used only in defense projects, it became a commercial product in 1948. It was mated with the electric frying pan in 1961 and named the "Happy Pan" at about the same time that NASA started using it for a host of applications from space suits to nose cones. Velcro was invented in 1948 by George deMestral. This Swiss engineer, who got cockleburs caught in his heavy wool stockings, saw the principle of tiny hooks and loops at work and reproduced the effect in woven nylon. The name came from a blend of *velvet* and *crochet,* which is French for *hook.* NASA has always used it, and each space shuttle contains about 10,000 square inches of Astro Velcro, which is a special type of the stuff. Tang, a commercial product of General Foods initially developed for the Army for prepackaged field rations, was bought off the supermarket shelf by NASA for the Apollo astronauts, who consumed it on the Moon. General Foods made much of this fact in its advertising.

tion enables doctors at city hospitals to communicate with paramedics in remote areas. Inertial guidance systems, essential to the navigation of spacecraft, are now standard equipment in commercial aviation and are used along with telemetry to track hurricanes. Still another example is remote sensing from orbiting Landsat satellites, which has been applied to such diverse jobs as monitoring oil spills, making inventories of forests, controlling water pollution, and prospecting for mineral deposits.

There is also a more basic level to space technology. Unlike mainline programs in such areas as Earth imaging, communications, and aeronautics—which are aimed directly at Earth application—there are other discoveries that are taken and reapplied by corporations, universities, and individual entrepreneurs. NASA often refers to this as "technology twice used." Ingestible, foamless toothpaste—developed for astronauts in a zero-gravity environment where spitting and frothing present a host of housekeeping problems—now is marketed commercially as NASAdent. It is especially useful for total-care nursing patients (who can choke on air bubbles), hospital patients, and others who are not always near a basin, and young children who often swallow toothpaste.

There are also inorganic paints (which help coastal bridges resist corrosion), collapsible towers (for applications ranging from portable radar to rock concert acoustics), and air tank breathing systems for firefighters (based on breathing systems that were used on the Moon). Sputnik-era technology led to watch batteries, food sticks, reservations systems, police radios, robotic systems, measuring instruments, insulation material, heart rate monitors, high-temperature lubricants, ceramic powders, solar panels, poison detectors, heated ski goggles, unscratchable sunglass lenses, Retin A to combat acne and skin wrinkling, water filters that attach to faucets, the Jarvik artificial heart, cordless tools, the liquid crystal wristwatch, freeze-dried food (first developed for John Glenn's 1962 Mercury orbits), and much more. Sports enthusiasts continue to benefit from the composite graphite fiber of tennis rackets, golf clubs, and fishing poles.

But the real value is in the large areas of new technology, which helped define the last years of the twentieth century and are poised to dominate the first decades of the twenty-first. NASA has had a program of technology development for satellite communications since the agency was established. The communications satellite industry, which began as a NASA program in 1961, became fully commercial in 1965; within twenty years it had become a $3-billion-a-year business, directly and indirectly employing more than one million people worldwide. It has grown geometrically since 1985. NASA is quick to point out that the growth period in this field is far from over. The beneficiaries have ranged from new and old communications giants to hundreds of smaller companies that did not even exist when the first Telstar communications satellite went into orbit. The communications satellite is the cornerstone of the wireless revolution now in progress.

Sputnik.com

The impact of Sputnik was also felt through the institutions it created. In 1958 the newly formed Advanced Research Projects Agency (ARPA) was composed of a far-flung pool of scientists, laboratories, and consulting firms. They needed a network that would allow the sharing of mainframe computers, which were very expensive and outside the reach of many members of the pool. In 1968 ARPA awarded its first contracts for the ARPANET, the forerunner of the Internet. ARPANET's physical network was constructed in 1969, linking four stations: UCLA, Stanford Research Institute, the University of California at Santa Barbara, and the University of Utah.

The first of these four "nodes" was a tall, gray box installed on September 2, 1969. A small crowd gathered inside Professor Len Kleinrock's lab at UCLA to watch flashing white lights as bits of information silently flowed along a fifteen-foot cable between two bulky computers. It was a test of the technology that remains at the foundation of the Internet. Then, on October 20, a group of computer scientists at UCLA again made history by getting their computer to "talk" to another at the Stanford Research Institute in northern California.*

By 1971 there were fifteen nodes, and in 1972 the first E-mail program was created by Ray Tomlinson of Bolt Beranak & Newman, a Cambridge,

*In 1999, UCLA professor Leonard Kleinrock, now sixty-five and credited as one of the founders of the Internet, recalled in an Associated Press interview: "We had a guy [Charley Kline] sitting at the computer console at UCLA wearing a telephone headset and a microphone, talking to another guy at Stanford. When everything was set up he was going to type the word 'log' and the Stanford computer would automatically add 'in' to complete the word 'login.' So our guy typed the 'L' and asked his counterpart at Stanford, 'Did you get the 'L'" and Stanford replied, 'Got the 'L.'" Then they did the same for 'O,' and then the whole system crashed!" He also told Associated Press science writer Matthew Fordahl that the first message ever sent from one computer to another was symbolic. "Put it into phonetics and you get (h)'ello, which is really quite appropriate."

Massachusetts, consulting firm. (Tomlinson also hit on the idea of the @, or "at," sign for E-mail addresses.) Overnight, electronic mail was flying in various networks. There was still a problem, however, as people could communicate within separate networks, of which there were several, but not between networks.

In 1962 Paul Baran, an immigrant from eastern Europe working for the RAND Corporation, was asked by the Air Force to determine how it could maintain control over its missiles and bombers after a nuclear attack. This was to be a secret military network that could survive a nuclear strike, decentralized so that if any targets in the United States were destroyed, the military would control nuclear arms through the network for a counterattack. Baran described several ways in which this could be done, but his final recommendation was for a "store and forward system" in which data are broken down into fragments, or packets, each labeled to indicate its origin and destination. The network would forward one of these packets from one computer to another until it arrived at its final destination, where it would be decoded and reassembled with the other packets. Baran's system was crucial to the realization of computer networks.

In 1974 Vinton Cerf from Stanford and Bob Kahn from DARPA—ARPA became the Defense Advanced Research Projects Agency in 1972—published "A Protocol for Packet Network Internetworking." It contained the design specifications for a Transmission Control Program (TCP) that would create a universal language, or "protocol," permitting diverse computer networks to interconnect and communicate with each other.*

DARPA's Bob Taylor networked all of the DARPA network's computers together for a total cost of $1 million, later labeled "the best million bucks ever spent by the Federal government." As worries about the Cold War subsided, the Pentagon let others into the system. In 1991 the government started to let businesses set up home pages on the toll-free system. At first there was disbelief that Web sites were free for the taking, and that there was no postage for E-mail, but this incredulity soon subsided and the rush was on to sign up. Within two years the number of commercial users outnumbered the academic ones. In 1995, the U.S. government got out of the business entirely, turning it over to Internet users. Early in the year 2000 it was determined that 52 percent of all American households were connected

*The term *internet* was short for the "internetworking" in their title. Cerf and Kahn were the first to use the clipped form "internet," in a paper on Transmission Control Protocol.

to the Web. During the year 2000 it was estimated that there were already twenty-five nations in which 10 percent or more of the population was on the Internet—including 136 million people in the United States, 27 million in Japan, 18 million in the United Kingdom, 16 million in China, and 6.6 million in Russia. At the end of 1999 there were 259 million users worldwide and, by the end of 2000, there were 400 million, which the Computer Industry Almanac estimated would jump to 1 billion by the end of 2005.

On September 2, 1999, on the thirtieth anniversary of the first ARPANET connection, the handful of pioneers met at UCLA in front of that refrigerator-sized first "node" to celebrate what they had wrought. The team of graduate students that took the crucial first step of hooking a computer to a switch in 1969 included future Internet leaders like Vinton Cerf, who later helped create the Internet's common language, Professor Len Kleinrock, the late Jonathan Postel, who pioneered its address system, and Steve Crocker, who led the development of the predecessor to TCP.

"In those early days, Len Kleinrock and his colleagues couldn't possibly have foreseen that they were on the ground floor of one of the most life-altering innovations of this century," said UCLA chancellor Albert Carnesale. Cerf credited two events as pivotal to making the modern Internet a reality: the breakup of AT&T (which, in his opinion, would have held its progress in check to maximize profits) and the launch of Sputnik.

Epilogue

Between the end of the Second World War and through much of 1957, there was much speculation about how and when the Space Age would start. Western visionaries without exception assumed the United States would lead off, although there was the occasional caveat. The groundbreaking 1952 *Collier's* magazine series on space travel opened with an editorial declaring that the United States must immediately embark on a long-range development program to secure "space superiority" for the West. "If [the United States does] not, somebody else will."

Somebody else did.

On Friday, October 4, 1957, the Space Age was under way with the launch of the first Sputnik. It was not only one of the most stunning engineering achievements of all time but the most public—heard and seen around the Earth. Sputnik was a Cold War event of enormous impact that engendered fear and surprise among Americans. It drove home the fact that the USSR had the firepower to deliver intercontinental multistage ballistic missiles to U.S. soil and any other part of the world, and suggested that outer space would became a new arena for warfare—an idea that had its advocates among the military elements of both nations.

Just as Sputnik was seen as a crisis for U.S. national defense, it was taken as proof that American education was failing, American science was inferior, and the American spirit was flagging. But Sputnik also reenergized America and ultimately gave it the upper hand in a high-stakes thermonuclear poker game. Sputnik gave President Eisenhower his "open skies," paving the way for reconnaissance satellites that later played a major role in the U.S. commitment to nuclear disarmament. Eisenhower was not the do-nothing president he often has been portrayed to be. Instead he was the quiet unsung hero of the Sputnik crisis, calmly leading the nation through a period of intense uncertainty, Cold War escalation, and rancorous rivalry among branches of his own armed forces.

The space race sparked by Sputnik created an intellectually competitive environment in the USSR and the United States that produced monumental advances central to modern science and technology—including the wired, wireless, and satellite-linked technologies of everyday life in the twenty-first century. Not only did Sputnik shift the balance of power between the United States and the USSR, it forever altered America's cultural and political landscape.

The United States could have put the world's first satellite in orbit in 1956 with a Jupiter rocket that reached an altitude of 700 miles. However, those working with the launch vehicle were ordered to make sure the fourth stage was a dummy. If the United States had put a satellite up first, American honor would have been assuaged at the beginning of the Space Age, but it might have led to a less benevolent contest in space or on the ground—and perhaps to a much different resolution of the Cold War—suggesting a definitive answer to a question posed to Wernher von Braun when he turned sixty in 1972. An old friend asked him, "Should there be any regrets?" He thought for a moment and then answered, "Perhaps it depends on whether it is better to be the first to put a machine into orbit or the first to put men on the Moon!"

Appendix: Sputnik's Long, Lexical Orbit

> The Russian word *Sputnik,* used for the satellite, briefly became the most famous word in the world.
>
> —Isaac Asimov, *Exploring the Earth and the Cosmos*

> We have kept up with the atomic bombs and the sputniks and all of that.
>
> —Kathryn (Mrs. George "Machine Gun") Kelly on being released with her seventy-year-old mother from a federal penitentiary after twenty-five years behind bars (quoted in the *Daily Oklahoman,* June 17, 1958)

Just as Sputnik was an important event, it was also an important word, one of those rarities that claimed its place in the English language—and other languages, for that matter—the very minute it was heard. It is, of course, only conjecture to say that the sound and look of the word added to its impact, but it is a word that did seem to carry its own shock value; completely the opposite of, say, *Salyut, Kosmos, Soyuz,* or the positively placid *Mir.*

The Russian name for the satellite was *Sputnik Zemlyi,* a term that means "traveling companion of the world," or Earth satellite. Almost at once *Sputnik Zemlyi* was shortened to *Sputnik,* and in that form it entered the languages of the world. The first syllable of the word was hardly ever pronounced as it would be in Russian. *"Spootnik! Spootnik!"* said a Russian woman I met recently. *"Sputt-nick* is an American word."

Indeed it was an American word. "Some new words take years to get into the language," said lexicographer Clarence L. Barnhart at the time. "She's a record-breaking word." He was so sure of it that twenty-four hours after the launch of Sputnik I he called his printer to dictate a definition to be added in the next edition of Thorndike-Barnhart's *Comprehensive Desk Dictionary.* Barnhart and other dictionary makers defined the word generically as meaning "Earth satellite."

While the early Sputniks were still in orbit, all sorts of things were done with the word. It was clipped by headline writers, as in "Sput 2: No Surprise," nicknamed "Sputty," and turned into a verb. For instance, the sudden popularity of space toys for

Christmas had people talking about their homes being *Sputnikked* for the holidays. It was turned into a slang expression, as in, to "go Sputnik," which was an early version of to "go into orbit." It was even used as a marketing buzzword: "Try our oranges—they are sputnik (Russian for 'Out of this world')," claimed a January 1958 magazine ad. It was blended with other words; *sputnik* plus *sorcery* equaled *sputnickery,* which made its debut in the *New York Times* in 1958. There were other odd suffixes, as in the case of *sputnikitis,* an unhealthy obsession with the satellite.

Puckish definitions purported to explain what the word *Sputnik* really means in Russian, such as "no tax cut for the Americans next year." *Sputnik* was even used as a term of deprecation, as when segregationist governor Orval E. Faubus was described as "that sputtering sputnik from the Ozarks" by Governor Theodore R. McKeldin of Maryland.

The second syllable became the all-purpose suffix du jour, attracting all sorts of odd and playful coinages. Sputnik I was "Propnik" to David O. Woodbury, author of *Around the World in 90 Minutes,* who suggested that it was more of a propaganda stunt than a technological feat. With the dog Laika aboard, Sputnik II was "Muttnik," "Poochnik," "Whuffnik," and "Dognik." The inability of America to get its satellite aloft was known as "Bottlenik." The failed American Vanguard was "Flopnik," "Dudnik," "Puffnik," "Pfftnik," "Sputternik," and "Oops-nik." *Time* (December 16, 1957) said of the Vanguard disaster, "Samnik is kaputnik."* When skyward-looking Americans started seeing mirages and other objects in the sky, they were written off as "whatnik" sightings.

While the event itself was still reverberating, the shock value of the word soon evaporated, and it became, in the light of continued Soviet success and U.S. failure, self-deprecating. The United States was now the "buttnik" of many jokes—for example: "Did you hear that America's scientists have invented a new Sputnik cocktail? It's two parts vodka and one part sour grapes."

The word was used in all sorts of atrocious puns and bizarre constructions. Vying for the worst pun was *saintnik* from a jolly *New York Times* Christmas 1957 editorial that contemplated the notion of having Santa's sleigh pulled by sputniks rather than reindeer. But there were countless runners-up, including *snuffnik* (stillborn in the issue of *Time,* November 11, 1957) for a Soviet medicine for Asian flu to be taken nasally. When the Russians in 1958 claimed a speed record for their TU-114 jet, it was called *speednik.* In 1959, the United States sent a replica of a typical suburban house to a Moscow trade exhibit—the modern home was split down the middle so that the Russians could witness American domestic life, and it was immediately called *splitnik.*

Most of these terms were ephemeral, and there is some doubt as to whether a few were ever actually used as claimed. For instance, *smoochnik* was presented in the February 9, 1958, *American Weekly* as a "kissing date" in the teenage vernacular. Others stuck: A golfer's *sputnik*—an extremely high drive off a tee—had a long life on the links.

Sputnik was promptly hyphenated, and remains so. NASA historian Roger D. Launius noted that "almost immediately, two phrases entered the American lexicon to define time, 'pre-Sputnik' and 'post-Sputnik.' It was hastily used as a modifier for

*One series of spacecraft with a *-nik* ending that had nothing to do with Sputnik were the Canadian *Anik* communications satellites of the 1970s. The name was chosen in a nationwide contest and is the Inuit word for *brother.*

words like crisis, diplomacy, shock and debacle which are still used. The legislative session which dealt with the crisis was known for many years as the 'Sputnik Congress.' "

Perhaps the strongest survival is in *beatnik*, created by *San Francisco Chronicle* columnist Herb Caen in his column of April 2, 1958, about a party for "50 beatniks." Caen later explained: "I coined the word 'beatnik' simply because Russia's Sputnik was aloft at the time and the word popped out." In his book on American youth slang, *Flappers 2 Rappers*, Tom Dalzell wrote, "*Beatnik* must be considered one of the most successful intentionally coined slang terms in the realm of 20th century American English."

Sputnik/beatnik led to a host of variations, including *neatnik* (someone who is well dressed and well groomed), *Vietnik* (someone opposed to the war in Vietnam), and *peacenik* (someone who is antiwar). In the 1970s the suffix was sent back to Russia attached to an English word: *Refusenik* referred to a Soviet citizen, often Jewish, who was refused permission to emigrate.

As the original English meaning of *Sputnik* (a generic English term for any unmanned Earth satellite) began to lose ground about the time that America's named satellites began to succeed, the name started to be used in nicknames, sobriquets, slang, allusions, and more. A singer who made a rapid rise to fame might be referred to as a "Sputnik tenor"; the British humor magazine *Punch* called Frank Sinatra, Sammy Davis Jr., and company "Hollywood sputniks," presumably because they were always in orbit. Bill Gold of the *Washington Post* noted that bureaucrats were sometimes called sputniks because "they ran around in circles." David Koresch, who later died as the leader of the Branch Davidians in Waco, Texas, was nicknamed Sputnik as a child because he was hyperactive. The nickname of the NBA's Anthony "Spud" Webb was given to him as a child and was a shortened version of Sputnik; Lenny Moore, of the Baltimore Colts, was also nicknamed Sputnik, so that to this day there are old-time Baltimoreans who think football when the word *sputnik* is uttered. *Sputnik* is still alive and well as a sports simile. In a 1995 Washington *Times* feature on a National Football League punter who routinely put the ball into orbit, it was noted that "[Reggie] Roby's sputnik-like shots have inspired awe among fans, teammates and opponents alike."

The Russian satellite inspired the names of all sorts of things. A Milwaukee bakery advertised doughnuts as sputnuts, the bun on the sputnikburger was topped with a pickle in orbit on the end of a toothpick, and one of the toys rushed to the market for Christmas 1957 was pednik, a revolving toy space vehicle powered by a foot pedal. At a hairdressers convention in Lansing, Michigan, on October 7, the Sputnik hairdo was born. The subject's hair was wrapped around her head with an upward flair, covering her ears, and was topped with a four-inch plastic model satellite complete with antennae. "The Sputnik Dance" was a minor hit record in 1958 from the Equadors, and an entrepreneur with an eye to the sky produced the Sputnik Fly Killer—two little pesticide-laden sputniks that hung from the ceiling and were guaranteed to kill flies. At one time or another Sputnik was a Norwegian folksinger, a ballet, a Ukrainian travel agency, a major Swedish horse race, a Russian forestry holding company, and a take-out double in the game of bridge.

One of the earliest adoptions was for a rock 'n' roll band whose business cards were spread all over Washington, D.C., at the height of the Sputnik scare. The card looked something like this:

Later, the name would be taken by a Scandinavian rock group that performed in tacky space suits. It also found new life in the world of gangs and punk rock bands, underscoring the fact that the word still retains a little of its old shock value. The English group Sigue Sigue Sputnik got its name from an article in the *International Herald Tribune* about a Russian street gang of that name. On the group's Web page, one of its members provides this rambling explanation: "Somehow [the name] captures the essence of the band. . . . The idea of this Fagin-like group of money launderers in Moscow is so right. . . . The name makes a great story every time it's printed. . . . Like all great names . . . [it's] so different. . . . Even the Sputnik connection is great . . . man's first object in space . . . like some kind of man-made god." The punk rocker is thrilled to find "Sputnik is [also] the name of a Manila motorcycle gang who when they kill people cut a notch in their arm."

The name shows up as a modifier in some unlikely places. The *Sputnik Christ* is the popular name for a mass-produced print from a late-1950s painting by Warner E. Sallman depicting a monumental Jesus standing atop a mountain with his arms extending into outer space, as planets, meteors, and satellites whiz by his head. The "Sputnik-weed" is the popular name of an invasive, one-celled algae that attaches itself to the back of oysters and whose single cell may extend for a mile. It got its name when it showed up in Long Island in 1958 and reminded people of the Sputnik's trailing antennae.

Sputnik was also the generic name for and basis of a hot style of design. This style has become popular again among collectors of 1950s modern furniture, including pieces designed by Charles and Ray Eames and accessories like the lava lamp. Sputnik-inspired objects became common in the late 1950s. One outrageous item was a hanging sputnik lamp, which had a round chrome center and eighteen arms that reached out to bulbs shaped like stars. The bulbs for these lamps were and are still known as "sputnik bulbs." Today on the eBay Internet auction site and elsewhere, there is a market for Sputnik design, as exemplified by spiked Christmas ornaments (originally used for futuristic aluminum Christmas trees of the early Space Age), Sputnik-inspired radio antennae, and the aforementioned lamps, including not only originals but replicas and variants such as the twisted Sputnik lamp, a table version, and a floor lamp that sports six illuminated spheres sprouting from a tulip pedestal.

For some, Sputnik eventually became both a crude unit of size and a measurement of time—even into the 1990s. Amy M. Spindler, writing in the *New York Times* of October 1, 1995, said, "Karl Lagerfeld taught the public to expect the unexpected. Collections swung from long to short, shifted with the popular music of the moment, were covered with exaggerated chains or pearls and topped by Sputnik-size hats." For a while in the early 1990s, one way sportswriters used to say that a team was getting old was to say "most of these guys are older than Sputnik," meaning they were born before the first launch, just as back in 1957 something could be "as new as Sputnik." The 125th anniversary of Frank Lloyd Wright's birth was noted in the *Arizona Republic* of May 17, 1992: "Wright was born in 1867, two years after the end of the Civil War, and died April 9, 1959, two years after the launching of Sputnik."

But perhaps the longest-lasting use to which the term *Sputnik* has been put in America is as a metaphor for a national challenge. "It is time for American policymakers to meet the challenge posed by the new global information economy much the way an earlier generation responded when the Russians launched Sputnik" is how Senator Larry Pressler, chairman of the Commerce Committee, addressed a 1995 plan to overhaul the nation's communications laws. The metaphor seemed to get more powerful over time. In the United States, every real or imagined weakness in the fabric of national life called for a Sputnik-like mobilization of national resources—or, as it is sometimes stated, "a new Sputnik." The Sputnik for 1999 was the anticlimactic Y2K challenge. Sometimes the thought was framed in terms of "What this country needs is a new Sputnik."

Perhaps the ultimate irony is the extent to which the term has become symbolic of faded Soviet glory and Russian decline. "There are the open manholes, the broken pavement," said an American Peace Corps volunteer in Moscow to a *Baltimore Sun* reporter in 1994, "in a country that built a sputnik before we did. Yet they don't care enough to do anything for people. It's idiotic and insane."

Eventually, the name and the very idea of a Russian Sputnik lost its ability to turn heads even when given symbolic meaning. A post-Soviet Sputnik fired in friendship was of so little import that many, if not most, American newspapers did not even bother to run this small story released by the Russians: "A space capsule containing a cut-glass replica of the Statue of Liberty and greetings from Russian President Boris Yeltsin has splashed down off the Washington coast on November 23, 1992. The Resource 500 Sputnik, which had circled the earth 111 times in six days, was billed as Russia's first private space launch and was targeted at the United States as a peaceful promotion marking the end of the Cold War."

A LITTLE ROCKET SCIENCE

Although the word *Sputnik* has fallen on hard times, another term of the Sputnik era is still very much with us: *rocket scientist*. In theory, it refers to any scientist or engineer working to put objects into space, but it is seldom applied in that sense. The term is commonly used as a measure of intelligence, or lack thereof, as in the construction "You don't have to be a rocket scientist to . . ." The term saw post-Sputnik use as early as October 10, 1957, in a *New York Times* reference to the Army team in Huntsville.* Later Wernher von Braun was often identified as America's leading rocket scientist—even though he and most of those given this label were actually engineers. As for its origin, the term probably came from an oft-quoted full-page ad run in the *New York Times*

*The term *rocket scientist* had been used as early as 1954 to refer to Wernher von Braun and members of his Rocket Team, as in "The Rocket Scientists Settle Down," an article by Hodding Carter on the Germans in Huntsville in the November 12 issue of *Collier's*.

in 1957. This ad reproduced an October 15 editorial in *Missiles and Rockets* admonishing President Eisenhower. It ended with the line "You don't have to be a scientist, Mr. President, to solve this problem. You must be a leader." Although the word *rocket* did not appear in this sentence, the editorial was about rockets, and so this is generally taken to be the first appearance of *rocket scientist* in the sense of a measure of IQ.

The closely related term *rocket science* has been used in such pedestrian statements as "Learning to tie your shoelaces ain't exactly rocket science."

Author's Note, Acknowledgments, and Dedication

In a very real sense, I am an eyewitness to some of Sputnik's most memorable influences on the West. As a teen, I watched Sputnik, enthralled by the adventure of the space race. As a young man, I was a Cold Warrior forced into uniform by the building of the Berlin Wall, and I was stationed on an aircraft carrier based in Mayport on the east coast of Florida that helped to recover U.S. astronauts from splashdown.

Later, I worked as a reporter covering the Gemini and Apollo missions for *Electronics* magazine. I have long collected material on Sputnik's impact on realms as diverse as industrial design and civil rights.

My first article on Sputnik appeared in 1982, and since then I have been obsessed with the subject. Many individuals have tolerated this obsession, and I thank those who actually helped me fuel it. Five who helped me with that original article were the late Homer E. Newell, Joe McRoberts, John Townsend Jr., Lee D. Saegesser, NASA archivist for more than thirty years, and Jack Limpert of the *Washingtonian,* who published it.

I would like to give thanks to Nadine Andreassen, Colin Fries, Steve Garber, Mark Kahn, Jane H. Odom, and chief historian Roger D. Launius at the National Aeronautics and Space Administration History Office in Washington, D.C., who gave me important advice during all phases of the project, including a review of the final draft. Other NASA helpmates include David R. Williams of the Goddard Space Flight Center; Mike Wright, historian at the Marshall Space Flight Center, Huntsville, Alabama; Barbara E. Green, Lisa Fowler, and Elaine E. Liston at the Kennedy Space Center; Alan Ladwig, both as the senior adviser to the administrator at NASA and in later incarnations; and Leonard David of Space.com.

At the Smithsonian's National Air and Space Museum, Michael J. Neufeld shared his time, files, and an early copy of his article "Orbiter, Overflight, and the First Satellite: New Light on the Vanguard Decision," and Tom Crouch provided invaluable assistance.

I would also like to thank Mike Baker at the Redstone Arsenal; Stephen Bates of the *Woodrow Wilson Quarterly;* Gini Blodgett, research manager at the National Press Club Library; Mark C. Cleary, historian at Patrick Air Force Base, Florida; Dwayne A. Day of the Space Policy Institute; George M. House, curator at the Space Center, Alamogordo, New Mexico; Dave Kelly at the Library of Congress; Rick Sturdevant, historian with the Air Force Space Command; John Taylor at the National Archives; Robert N. Turner of the Space Center, Alamogordo, New Mexico; John Vernon at the National Archives; and Charles P. Vick of the Federation of American Scientists. Thanks also to Benedict Fitzgerald for his recollections; Steve Hash for bringing new life to old photographs in his magic darkroom; Don W. Savage for his fine-artist efforts on behalf of this book; Jim Hoekje for his recollections of his 1956 Vanguard work and the October surprise in Washington; Brandi G. Horne at the Economic Development Commission of Florida's space coast; bookseller Richard H. Jackson for his recollections of SCORE; Tom Pitoniak for some good word work; Milt Salamon at *Florida Today,* whose columns keep track of the veterans of the early days of the space race; John Driscoll of Wesleyan University for setting up interviews in Middletown with Bob Rosenbaum and Duffy White; and Terry Souvain of the Senate Appropriations Committee. Ronald L. Potts, consultant, Naval Space Technology Program, and his associate Reid Mayo furnished insights into the pre-Sputnik Naval Research Laboratory and the IGY. Henry Allen offered encouragement and insight at a moment when it made a difference, and Professor Kim McQuaid of Lake Erie College—one of the hottest Cold War historians—provided a great and totally impromptu "consult," which turned into a full-tilt vetting of the manuscript and many important suggestions for improvement.

Over the course of several years, many friends and fellow writers helped with this project. I am deeply indebted to Alan Austin for letting me work some of the kinks out over Chinese food, Kai Bird for sharing leads and references, Percy Dean, who, among other things, photographed Sputnik II, and Jack Waugaman for lending me magazines and newspapers of the period from his collection.

For more than five years Douglas E. Evelyn, Peter Gibbon, Bob Skole, and Bill Hickman were valued sets of eyes and ears for this project, finding Sputnik leads in all sorts of odd but useful places. I would also like to thank the late John Masterman; Frederick A. Koomanoff for his valued insights; Brad Whipple for his recollections; Tim Wells for his knowledge of the period; and Ned Dolan for his many leads and insights into the covert world created, in part, by Sputnik.

Thanks also to John M. Morse for connecting a few important dots and James Srodes for his contacts, leads, book loans, and advice. At the National Press Club in Washington, thanks go to Stanley Hamilton for a number of things (including the all-important connection between Sputnik and Mrs. "Machine Gun" Kelly), Ross Mark for his recollections of the earliest American launches, and Hale Montgomery for establishing important links. Special thanks go to Press Club colleague Paul Means, who was there from the beginning, for his unswerving help and an important reading of an early version of the manuscript. James Ludlum and his crew of book scouts at the Friends of the Library Book Sale in Wheaton, Maryland, helped me find elusive, ephemeral documents on the early days of space. Chuck Rich's assistance included finding a tape of the late Mickey Katz's contribution to Sputnik culture (and Western civilization): the song "Nudnik, the Flying Schissel." (*Schissel* is Yiddish for *bowl* or dish.) Bill Hines, a journalist from the days I covered NASA, provided insight into the early days of space.

Thanks to everybody at Walker & Company, especially Marlene Tungseth and Vicki Haire. Special thanks go to George L. Gibson for his unflagging support and Jackie Johnson for applying her great talent and good humor to getting this story right. Finally, my great appreciation goes to my long-time agent, Jonathan Dolger, for all of his support.

Thanks of the highest order also go to Joseph C. Goulden for leads, book loans, wonderful discoveries, and his enthusiasm for this project, as well as to Nancy Dickson, who got me through many early dysfunctional drafts and who gave of her time so I would have time to write this book.

Finally, the late Isabelle C. Dickson, my mother, is in large part responsible for this book. She was the consummate space enthusiast who never missed a televised launch and who read everything she could on the subject. Even when my interest flagged, she always wanted to talk about space: what was next, what it was all about. She belonged to a current events club—the Up-to-Date Club—where she and other women delivered talks on topics that they had taken on as an avocation. My mother dabbled in other subjects, but she always came back to space. She claimed it as her first love.

In September 1999, a year after her death, I got around to unpacking the box of space papers that she had carefully assembled over the years. The box contained such items as the first prayer from space, which hung over her desk for years, and a small pile of notes she had taken on the weekend of the first Moon landing. Her bibliography "Astronomy, Satellites and Space" was useful in preparing this book.

If nothing else, the box of space oddments underscored the span of the one true outside-of-the family adventure that had helped define her rich life. Like others, she had been mesmerized by the photographic images coming back from the Moon and the potential for using satellite images to better life on Earth.

In one of her papers she asked, "Looking out from the moon desert did our planet, floating in deep space, appear like an oasis, beautiful, perfect and with no room for hunger, war or potential strife?"

Another of her papers on space, which she delivered in 1972, contained this line: "Now, fifteen years later, it is difficult to recall the devastating effect of Sputnik I."

Now, more than forty years later, I have tried to give that "devastating effect" its due.

This book is dedicated to her.

Notes

Many sources are referenced in the body of the text and in footnotes. The following notes are meant to complement those embedded citations and to provide a forum for a dozen or so digressions outside the narrative scope of the story. For full publication data on works cited in the notes, see the bibliography.

INTRODUCTION

The new, previously secret, material alluded to in the introduction came into sharp focus after February 22, 1995, when President Bill Clinton signed Executive Order 12951 on the "Release of Imagery Acquired by Space-Based National Intelligence Reconnaissance Systems." With new documentation and revelations, old assumptions and bits of conventional wisdom were blown away, and different answers have been suggested for such questions as: Could the United States actually have gone into space first, and if so, what held it back? How close did a faction within the military come to "accidentally" putting a satellite in orbit months before Sputnik? A number of references were used to answer these questions, including Day's several articles focusing on the period ("A Strategy for Space"), the work of Hall (e.g., "The Eisenhower Administration and the Cold War"), Neufeld (e.g., his essay in Launius et al., eds., *Reconsidering Sputnik*), Launius ("Eisenhower, Sputnik, and the Creation of NASA"), as well as works published by the Central Intelligence Agency (Pedlow and Welzenbach's *The CIA and the U-2 Program, 1954–1974* and Ruffner, *Corona,* to name two).

Those of us who were children in the years before Sputnik lived with the fear of polio until 1953, when Jonas Salk found a cure. This dreaded disease crippled and killed many people, mostly children. In the United States in 1951 alone, it struck more than 50,000, killed 3,000, and left thousands crippled. Also known as infantile paralysis, it was a terrifying epidemic, especially during the summers of 1942 through 1953. Careful parents, like mine, kept their children out of movie theaters, ballparks, and public swimming pools, where the virus was believed to lurk. During peak outbreaks, there were mass closings of any places where children congregated in large numbers. In 1953 Jonas Salk discovered the polio vaccine and became a hero of the first order. Thanks to him, summer had been returned to millions of children—along with the gift of optimism and a belief in the redeeming power of science, problem-solving, and intellect.

Franklin R. Chang-Dìaz told of his Sputnik-inspired career at the conference "Space Exploration at the Millennium" held on March 24, 1999, at American University in Washington, D.C. What most of us who saw Sputnik in orbit would not learn for decades was that we were probably seeing the orbiting booster and not the satellite itself. The effect was the same.

ONE. SPUTNIK NIGHT

The placement of individuals on Sputnik night and descriptions of the party at the Soviet Embassy appear in news accounts of the time and in the individual biographies in Thomas (*Men of Space*). Events in Huntsville are chronicled in the local newspapers, by Medaris and Gordon (*Countdown for Decision*), and in the papers in the Medaris archives. Events in Cambridge, Massachusetts, are described in an essay by E. Nelson Hayes (in Clarke, ed., *The Coming of the Space Age*). Presidential recollections of Sputnik night appear in Johnson (*The Vantage Point*) and Eisenhower (*The White House Years*). The Kennedy placement is made in Durant (*AAS History*, vol. 3, *Between Sputnik and the Shuttle*), the information on Michener is in Reed ("The Day the Rocket Died"), and the astronauts' recollections come from their various autobiographies (e.g., Shepard and Slayton's *Moon Shot*). Ross Perot, Konstantin Feokistov, and German Titov were interviewed on May 16, 1997, at the National Air and Space Museum in Washington. Goldin's recollection was given in an impromptu 1995 interview as well as in several of his speeches (e.g., "For the Benefit of All Humankind" to the American Medical Association, October 20, 1992). The recollections of Sergei Khrushchev on his father's reaction to Sputnik appear in the essay "The First Earth Satellite," which is included in Launius et al., eds., *Reconsidering Sputnik*. This supersedes earlier impressions that Khrushchev was somewhat less involved.

The section titled "The Press Reacts" is almost entirely based on a large chronological collection of press clippings at the NASA History Office in Washington, D.C. The collection, which extends from the early part of the twentieth century to the early 1960s, was begun by Lee Saegasser and completed by Mark Kahn. It was augmented by a similar scrapbook collected by General Medaris and available at his archives at Florida Tech.

TWO. GRAVITY FIGHTERS

An excellent lay explanation of Newton's laws appears in Park (*Voodoo Science*).

The original story of the Brick Moon appears in *Atlantic Monthly* 24 (Oct., Nov., Dec. 1869): 451–60, 603–11, 679–88. Also published in Hale, *The Brick Moon and Other Stories* (New York, 1899). The full text of the 1899 book is in John M. Logsdon, ed., with Linda J. Lear, Jannelle Warren Findley, Ray A. Williamson, and Dwayne A. Day, *Exploring the Unknown: Selected Documents in the History of the U.S. Civil Space Program*, vol. 1, *Organizing for Exploration* (Washington, D.C.: NASA 1995), NASA SP-4407. Some people took the story seriously. Seven years after it appeared, Hale got a letter from Asaph Hall, an astronomer at the Naval Observatory in Washington, D.C., reporting that he had discovered that Mars had two small moons, one of which revolved around the planet more swiftly than Mars turned on its axis. According to Hall, "The smaller of these moons is the veritable Brick Moon." The gesture warmed Hale's heart and he wrote, "That, in the moment of triumph for the greatest astronomical discovery of a generation, Dr. Hall should have time or thought to give to my little parable—this was praise indeed." Sir Arthur C. Clarke's comment on the Brick Moon was made in a letter that appeared in *IEEE Spectrum* for March 1994.

In 1945, long before the invention of the transistor, Clarke wrote an article for the magazine *Wireless World* outlining how global communications could be provided using three satellites positioned evenly around the equator at an altitude of 26,000 miles. Clarke was slightly off in calculating the correct altitude, which is 22,300 statute miles, but his theory was dead on. Satellites would orbit the Earth in exactly one day; therefore they would appear to be stationary to a person on the Earth. Clarke's geostationary satellites would relay signals from one point on the Earth to another and contrast to the Earth-circling satellite, which would, for different purposes, orbit the Earth in a matter of minutes.

The history of rocketry appears in a number of sources, such as Canby *(A History of Rockets and Space),* von Braun and Ordway *(History of Rocketry and Space Travel),* and Emme *(The History of Rocket Technology).* The lives of Tsiolkovsky, Goddard, and Oberth are well documented in the major histories of space, including Clarke *(The Exploration of Space),* Crouch *(Aiming for the Stars),* Heppenheimer (*Countdown*), and McDougall *(The Heavens and the Earth).* Even after two Sputniks, the (Encyclopedia) *Britannica Book of the Year* for 1958 included this statement: "Although the history of rockets is more than 2000 years old, modern rocketry dates from the researches of R. H. Goddard following World War I." Lehman *(This High Man)* is the best single source on the life of Goddard. Roger Launius, NASA historian, believes that a new biography is needed.

THREE. VENGEANCE ROCKET

German rocket clubs, von Braun, and the development of the V-2 appear in Neufeld *(The Rocket and the Reich)* and Crouch *(Aiming for the Stars).* Those willing to take a critical look at von Braun and the Rocket Team were few and far between for many years until the publication of Hunt's 1991 *Secret Agenda,* Neufeld's 1995 *The Rocket and the Reich,* Piszkiewicz's 1998 *Wernher von Braun,* and Crouch's 1999 *Aiming for the Stars.* An early issue of a sensational magazine called *Confidential* (the clipping in the NASA History Office lacks a date), which carried an article titled "The Ex-Nazi Who Runs Our Space Program," gave an unflinching portrayal of the V-2 labor conditions: "The slaves were in rags, they were kept at 500 calories a day, and when they fell ill, the adjacent Nordhausen concentration camp gassed them."

Olsen *(Aphrodite)* not only made the story of Aphrodite public and clear for the first time but added the chilling footnote to the tragic tale about the meeting in Dallas. The key Aphrodite documents are stored at the National Archives in Suitland, Maryland, and some can be viewed on its Web site, www.nara.gov/

Medaris's close call with one of von Braun's rockets is recounted in Medaris and Gordon *(Countdown for Decision).* "In a manner of speaking, I met Wernher von Braun—or at least was on the receiving end of a message from him—a dozen years before I was introduced to him personally." When that meeting did occur, it led to a partnership that would put America in space in 1958 and, by extension, put Americans on the Moon. Medaris would be as important to von Braun's future success as Oberth had been to his past.

Lasby *(Operation Paperclip)* and Bower *(The Paperclip Conspiracy)* describe the relocation of the Rocket Team to the United States. The Fort Bliss/White Sands period is detailed in Ordway and Sharpe *(The Rocket Team).* Von Braun and the Rocket Team arrived in America at the perfect moment. On September 26, 1945, a WAC-Corporal reached an altitude of 43.5 miles, a U.S. record at the time and the first liquid-fueled rocket developed with government funds in the United States. It would soon be married to the V-2 as the upper stage of a rocket in Project Bumper. Significantly for space science, the WAC-Corporal was the progenitor of a larger, improved sounding rocket, called Aerobee; later versions of the Aerobee were capable of carrying a substantial instrument load above 100 miles and became mainstays of the American high-altitude research program. Meanwhile, in 1946 the Naval Research Laboratory began work on the Viking, a large research rocket meant to replace the V-2. In the end, fourteen Vikings were launched between May 1949 and May 1957. They would figure prominently in the American journey into space.

The definitive work on Korolev is Harford *(Korolev),* who gave the West his full story. In the years following Sputnik the world became obsessed with trying to determine who this elusive "chief designer" was. At the time of his death in 1966, Stuart H. Loory reported from Moscow for the *New York Herald Tribune* that his identity was as closely guarded behind the Iron Curtain as it was beyond it. In the West the veil of secrecy worked so well that NASA administrator James E. Webb could not bring himself to send a note of condolence to his Soviet counterpart because, in part, "Korolev has never been known."

The RAND report of May 1946 was titled "Preliminary Design of an Experimental World-Circling Spaceship." Despite its initial rejection, the report helped establish the reputation of RAND, and, quoting a 1963 RAND history, "the findings were the foundation of a continuing program of research which, through the years, was to yield hundreds of reports on space technology, including meteorological satellites, planetary explorations, and ballistic and glide rockets."

FOUR. AN OPEN SKY

The Silver Spring genesis of the IGY is detailed in biographies of participants in Thomas *(Men of Space),* among other places. John Eisenhower *(Strictly Personal)* is key to understanding Dwight Eisenhower's discussions with the Russians on his "Open Skies" proposal. The prime source for following Slug, Orbiter, and the Stewart Committee is Neufeld's essay "Orbiter, Overflight, and the First Satellite," in Launius et al., eds., *Reconsidering Sputnik*. Neufeld also allowed me to see his supporting documents for this essay. Neufeld is working on a biography of von Braun, which should shed more light on this period. The all-important 1955 directive from the National Security Council that decreed that the U.S. program could not employ a launch vehicle currently intended for military purpose was titled "U.S. Scientific Satellite System" (NSC 5520) and was signed on May 26, 1955.

Medaris and others came to believe over time that it was the Rocket Team's Nazi past rather than something as simple as the fact that von Braun's plans did not fit in with the needs or timing of the International Geophysical Year. Von Braun's proposed satellites were shows of power rather than a scientific experiment. To this day, there are those who believe that it was pure anti-Germanism at work. I was told in 1999 by a historian working for the Army in Huntsville that America lost its chance to be first in space because Eisenhower did not like von Braun and the rest of the team. In the foreword to Green and Lomask *(Vanguard),* Charles A. Lindbergh asks the question why the Redstone was not picked for the space program, and after giving all of the usual reasons, he adds "that any conclusion drawn would be incomplete without taking into account the antagonism still existing toward von Braun and his co-workers because of their service on the German side of World War II." (Lindbergh's remarks are dated August 11, 1969, and are taken from NASA SP-4202, 1970, new p. 125.) The Redstone was pure and simple an American incarnation of the Nazi V-2, which had been used against an American ally. Eisenhower had fought the Nazis, and there were many in the American scientific and military elite who despised the swaggering cadre of recent enemies and their Teutonic egos. This was no petty prejudice but a reality: America was less than a generation removed from the tyranny of a self-styled master race threatening to conquer the Earth. If there was an image that shone through, it was the contrast between the self-effacing, professorial Hagen, who was in charge of Project Vanguard, and the bombastic ox-shouldered rocket salesman von Braun, who seemed bent on exploiting the military uses of space.

Even without going first, there were some moments when the Nazi past came to haunt the American space program. In 1960, when a movie about von Braun, *I Aim at the Stars,* was released, all sorts of anger was unleashed. The Belgians were angry because Antwerp had been destroyed by the V-2, the French were angry because von Braun had used French slave labor to build the V-2, and the British were stunned by the fact that Columbia Pictures would stage a gala London premiere in Leicester Square. Anti-Nazi pamphlets rained down on the select audience like so many V-2s, and two men got on the stage with a twelve-foot banner reading "Nazi Braun's V-2 Rockets Killed and Maimed 9,000 Londoners." Even the man who played Von Braun, Austrian Curt Jergens, was quoted as saying, "Von Braun is an unpleasant hero. I don't like him much."

In light of the information reported on in this book, it is clear that anti-Germanism was not the reason that the Army was not allowed to go into space first.

Sir Arthur C. Clarke's comment on the possibility of an accidental launching appears in his book *The Promise of Space,* and the story of the transcontinental speed record appears as a firsthand recollection in Glenn and Taylor *(John Glenn).*

The storage of the Jupiter C rockets in Huntsville and the Orbiter satellites at JPL is well documented in Thomas *(Men of Space),* Medaris and Gordon *(Countdown for Decision),* and in the Johnson hearings (U.S. Congress, Senate Committee on Armed Services, *Inquiry into Satellite and Missile Programs).*

The initial launch date for the American satellite was set for November 1957, early in the IGY, but it would slip. A signal moment came in 1956 when writer and visionary Sir Arthur C. Clarke came to Baltimore as chairman of the British Interplanetary Society to discuss, among other things, the coming importance of satellites for radio, television, astronomy, physics, global communications, and the remote sensing of Earth. As soon as Clarke arrived in Baltimore, he insisted on dropping by the Glenn L. Martin Company to look at the rocket selected to launch the world's first satellite. Vanguard was a low-priority project even with the prime contractor; Clarke was astonished to find that the company held the Vanguard in so little regard that its satellite rocket builders were assigned to "hot and tawdry offices above an aluminum smelting machine, beneath pigeon-filled rafters." Clarke's visit was recalled in an article in the *Baltimore Sun* on October 4, 1997, the fortieth anniversary of the first Sputnik launch. According to this article, the people Clarke met—including members of the Vanguard engineering team—seemed largely unaware of what was about to happen. "Wouldn't they be surprised," Clarke noted, when "in a year or so, the American satellites start racing overhead west to east—and the Russian ones, north to south."

To the American public, there was never a question as to who would be first in space: Americans knew that they would be. In fact, the United States was ready to go, and the assumption of American superiority had an official basis. The periodic reports on the American effort included occasional hints or mention of slight delays, but as far as the public was concerned, the program seemed to be moving along quite nicely. For instance, on September 22, 1956, the *New York Times* carried a detailed article describing not one but seven types of satellites being fabricated in the United States for launching within fifteen months. By most accounts, the big day would come in the spring of 1958.

Even if one were looking for evidence of the Russian preparations in American popular and scientific literature, one would be hard-pressed to find more than an occasional and invariably dismissively short reference. In Lloyd Mallan's 1955 work *Men, Rockets, and Space Rats,* which examines the upcoming drive into space, the only mention of a potential Russian satellite is in a quote by a scientist, who says that both the United States and the USSR have "sponsored" scientific satellite programs.

FIVE. THE BIRTH OF SPUTNIK

Events leading up to Sputnik in the USSR are recounted in Harford *(Korolev),* Romanov *(Spaceflight Designer),* Riabchikov *(Russians in Space),* and Oberg *(Red Star in Orbit).*

The memo sent to Berkner by Bardin was declassified in the year 2000 and presently resides in the IGY file in the NASA History Office in Washington, D.C., along with a copy of the Kotelnikov presentation. Many of the clippings in the aforementioned collection of press clippings at the NASA History Office, as well as a card file in the same office, provide evidence of the Soviets' eagerness to tell the world they were about to launch a satellite and of America's unwillingness to listen.

Why was the United States unable to accept Soviet science on the eve of the Sputnik launch? One explanation is that an incredible development in the Soviet Union during 1948 bolstered the notion that Soviet totalitarianism was totally at odds with Western science. This development was the total acceptance of a Lysenkoist idea of genetics. Trofim Lysenko, a Soviet botanist, challenged the traditional Mendelian account of genetics based on heredity by arguing that environment determined plant characteristics. He made his views known in an August 1948 address to the Soviet Academy of Natural Sciences, which resulted in a government decree that all Soviet researchers must accept Lysenkoist principles or face dismissal. In the West, scientists hooted derisively at this development, knowing the proven superiority of the Mendelian theory. After seeing the impact of Communist dogma on the way science was practiced in the Soviet Union, the highly autonomous American scientific community believed it had nothing to fear from Soviet science. The running gag was that "Soviet science" was an oxymoron.

The incident with Senator Ellender was recalled in 1972 on the occasion of Wernher von Braun's sixtieth birthday. Lieutenant General James M. Gavin was one of many who wrote letters to von Braun offering congratulations and reminiscing about his distinguished career. Gavin noted in his letter that as von Braun listened to Ellender, he nodded his head acknowledging the senator's remarks. "I was a bit concerned that the recorder of the hearings might record Von Braun's gesture as agreement," Gavin wrote. "So, I handed him a note suggesting that he be careful not to give the impression of agreeing with the Senator, since I knew that neither of us did agree with him. The Chairman of the committee brought the hearing to an end and then called me before him and threatened to throw me out of the room for attempting to influence a witness. It brought the hearing to an end and no one was convinced that the Soviets could launch a satellite." Gavin's letter appears with the other testimonials in a bound collection sent to von Braun; copies exist at both the NASA History Office in Washington, D.C., and the Kennedy Space Flight Center.

Bissel's comments appear in Bissell, Lewis, and Pudlo *(Reflections of a Cold Warrior),* along with this comment on the overflights: "One of the ironies of history was that months before the launching of Sputnik, an event that would shake American self-confidence, the United States was flying with impunity above this most secret of Soviet military locations." Day's article, as well as a subsequent paper (titled "Cover Stories and Hidden Agendas: Early American Space and National Security Policy") delivered at a 1997 NASA Symposium on Sputnik, showed that the Department of Defense was not interested in this strategy, but the Central Intelligence Agency was—explaining why the CIA contributed the money. Of all the many documents Day unearthed, perhaps the most telling was a memo from the National Security Council planning board dated November 9, 1956. This memo noted that "the USSR can be expected to attempt to launch its satellite before ours and to attempt to surpass our effort in every way. It is vitally important in terms of the stated prestige and psychological purpose that the United States make every effort to (1) make possible a successful launching as soon as practicable and (2) put on as effective an IGY scientific program as possible. The prestige and psychological setbacks inherent in a possible earlier success and larger satellite by the USSR could at least be partially offset by a more effective and complete scientific program by the United States. Even if the United States achieved the first successful launching and orbit, but the USSR put on a stronger Scientific program the United States could lose its initial advantage."

The undated "Olive Branch" memo was written after the beginning of the IGY and before the Sputnik launch in October. It was placed in the Sputnik folder after the event with all other pre-Sputnik documents relating to the Soviet satellite plans. It was declassified by the NASA History Office in Washington, D.C., in March 2000. The Potts and Hines responses to the memo were solicited by the author in December 2000. Being a good journalist, Hines asked if the memo could have been a fake. Several authorities within NASA, including Jane Odum, the archivist who released the package of documents in which it was embedded, say there is nothing about this document that is not authentic. Other material filed in the same jacket with the memo include one top-secret memo, which—despite the efforts of the author and the NASA archivist—had not been released at the time of this writing, when I was told it could take years.

Besides Harford *(Korolev),* direct quotes from Korolev come from Russian sources, notably Romanov *(Spaceflight Designer)* and Riabchikov (*Russians in Space;* originally published by Moscow's Novosti Agency in 1971 and quickly followed by the American Doubleday edition). The Romanov book was based on a series of interviews conducted with Korolev beginning four years before his death.

SIX. RED MONDAY

Once again, the NASA chronological clipping file proved invaluable to this rendering of the reported American reaction to Sputnik. Other helpful sources were the 1957 scrapbook in the Medaris archives, at the Florida Institute of Technology in Melbourne, Florida, Divine *(The Sputnik Challenge),* Bulkeley *(The Sputniks Crisis and Early United States Space Policy),* and Lule ("Roots of the Space Race").

Aside from containing a great collection of reactions, Lule's article provides insight into the press coverage and the degree to which it was provocative. An assistant professor of journalism at Lehigh University, Lule examined the language used by three members of the "prestige" press—*New York Times, St. Louis Post-Dispatch,* and *Los Angeles Times*—from October 4 to 31, 1957, the first four weeks of Sputnik's ninety-two-day flight. He found that Sputnik was portrayed in dramas "of defeat, of mortification, and of the dream and dread of space and that in the firestorm of criticism after Sputnik, the press played an incendiary role." He also concluded that the press marched to its own beat: "Although sources surely had influence, the press did resist many sources—including the President. In the final analysis, the press *selected* its primary themes, its main actors, its own dramas to enact."

Taking nothing away from Lule's disciplined study of news content, it was also evident that the rioters in the media were part of a larger mob, including many scientists who had no idea how sensational their speculative remarks would seem when they appeared in print. For example, Professor Fred Singer of the University of Maryland, in a quote in the *Washington Daily News,* proposed that the Moon be used as a testing ground for nuclear weapons, so that fallout would not pollute the Earth.

The Bounds letter was included in a file on Sputnik that was declassified in March 2000. The file is now available at the NASA History Office in Washington, D.C. Bounds, among others, was ascribing to the Russian satellite a role that America was planning for its own satellites. Many of the same Americans who had a low regard for Soviet technology on October 3, 1957, were imagining it to be much more sophisticated than it actually was. Despite periodic Soviet announcements to the effect that the beep of the satellite was nothing more than a steady signal to help trackers follow it, for days to come the suspicion continued that the satellite was collecting information and sending back ciphered reports. Others tried to guess what it held. In his 1958 book *Around the World in Ninety Minutes,* David O. Woodbury added a last-minute chapter on Sputnik, in which he speculated on what was hidden inside the "immense" Russian satellite. "There must have been 163 pounds of pay load unaccounted for and held secret. Instruments? Means of gaining information they do not wish to share with the world?"

Governor McKeldin was not the only Maryland politician to make news by invoking Sputnik. Maryland Republican state senator John Marshall Butler told the *Baltimore Sun* that future Sputniks should be shot down out of the skies as fast as they appear: "I would like to see our Armed Forces just say to them you put them up and we'll shoot them down. I think we'll come to that some day, and it's not too far off. The best way to deflate the propaganda value of Sputnik is to knock it out of the air."

The pressure that Eisenhower was under seems even greater over time as more information on this period becomes available. More than ever, the president needed loyalty from his top brass, not criticism. These were times of great trepidation for Eisenhower, who in 1957, according to documents made public September 2, 1998, in the *Washington Post*—Pincus and Lardner Jr. ("Eisenhower Issues Limited Nuclear Authority Absent Presidential Order")—granted authority to senior U.S. military commanders to retaliate with nuclear weapons if he was unable to respond to a nuclear attack by the Soviet Union. In other words, he had given his generals the right to wage thermonuclear war in the event that he was a casualty of the first strike.

The fact that the United States had gotten the most talented German rocketeers was well known at the time of Sputnik. Grigori A. Tokady, chief rocket scientist for the Soviet Air Force, recalled in *Stalin Means War* (1951) as quoted in Ordway and Sharpe (*The Rocket Team*): "The problem we face today is this: we have no leading V-2 experts. . . . We have no complete projects or materials of the V-2; we have captured no operational V-2s which could be launched right away. We have projects which may be very useful to us. We have the free or compelled co-operation of hundreds of German workers, technicians, and second-rate scientists, whose experience could be of value to us." Wernher Von Braun, who knew these facts firsthand, later told a Senate committee that Sputnik was a pure Russian effort "with little if any German assistance." Indeed, without the ironfisted, insanely jealous rule of Stalin, and the imprisonment of Korolev and others in prison camps, the Soviets might have gotten a satellite launched earlier. If they had grabbed Wernher von Braun and company, they would have been unstoppable. And if the von Braun team, which had been working on the V-2 since 1930, had gotten Adolf Hitler's approval to work on a satellite rather than weapons of vengeance, the first Earth-orbiting satellite could have had a swastika on it rather than a red star.

Washington was not pleased that the Army was giving Vanguard bad grades, especially when pitted against the Jupiter C. On the Monday after Sputnik, Army secretary Wilbur Brucker sent a memo to the secretary of defense. "Unfortunately," he said, "Project Vanguard does not appear to offer much hope of an early satellite. On the other hand, the re-entry test vehicle used in the Jupiter program, the Jupiter C, can be readily modified to provide an early U.S. satellite capability." Brucker also mentioned the three Jupiter Cs already fired: "Prior to the first launch of a JUPITER satellite, we can point out, if desirable for psychological purposes, that we already launched three satellite test vehicles."

At least thirteen members of the staff of the Naval Research Laboratory, including John P. Hagen, initialed a translation of Kotelnikov's remarks on the proper frequencies. The original translation and the routing slip were removed from a long-classified NRL Sputnik file in 1999. The file is available at the NASA History Office in Washington, D.C.

World public opinion surveys performed by the U.S. Information Agency are housed at the National Archives in Suitland along with CIA National Intelligence Estimates. American intelligence would, over the course of the next few years, give Sputnik an importance as a turning point and rallying cry that had initially eluded both the Kremlin and the White House. Joseph C. Goulden brought to my attention a CIA assessment dated November 4, 1960 (in Central Intelligence Agency, National Intelligence Estimate, "Main Trends in Soviet Capabilities and Policy, 1960–1965"), that put Sputnik into historic context: "Looking back to the weak and perilous position in which the new Communist regime found itself in 1917, remembering all the internal and external trials it has survived, and considering its growth in relative economic and military power over the last 20 years, the Soviet leaders are encouraged in their doctrinaire expectations about communism's inevitable triumph. That it was a Communist rocket which first ventured into space symbolizes for them that they are marching in the vanguard of history. They think they see a response to their doctrines and influence in the revolutionary turmoil of Asia, Africa, and Latin America. They expect to associate the people emerging from colonialism and backwardness with their own cause, mobilizing them against an ever more constricted world position of the Western states."

SEVEN. DOG DAYS

Radosh and Milton *(The Rosenberg File)* and Chang *(Thread of the Silkworm)* were the two most important sources for the early portion of this chapter. Laika's time of death—long a matter of speculation—was finally determined in Asif A. Siddiqi's book *Challenge to Apollo: The Soviet Union and the Space Race, 1945–1974*, published in 2000 by the NASA History Office. At least thirteen other Russian dogs were launched into orbit between November 1957 and March 1961. Five of the dogs died in flight. I interviewed Newell and Townsend for my article "How a Little Beep-Beep-Beep Scared America Half to Death," which appeared in *Washingtonian* magazine in September 1982. The suggestion that America was abuzz with thoughts of a lunar atomic bomb may seem outrageous, but it is not. Walter Sullivan, one of the most respected science writers in the country, wrote two stories for the *New York Times* about the possibility of a celebratory thermonuclear display, citing conjecture by "scientists in many lands." Sullivan and others were careful to mark this as a speculation but also reported on corroborative information such as strange radio signals, which suggested the story could be true. For its part, the Soviet Union through Tass had made it quite clear that Sputnik II was the celebration commemorating the fortieth anniversary of the great October Socialist revolution. The aforementioned chronological collection of press clippings at the NASA History Office in Washington, D.C., contains other examples of this fear of lunar nuclear attack.

The spark for the Johnson hearings was a memo with a link to Little Rock. On October 17, trusted adviser George E. Reedy prepared a memo for Johnson dwelling on the importance of Sputnik. The cover letter noted that Charley Brewton, one of the shrewdest political analysts Reedy knew, had flown all the way from Boston to Austin to help him with the memo. For Brewton, "the issue is one which, if properly handled, would blast the Republicans out of the water, unify the Democratic Party and elect you President." Reedy's memo began, not with Sputnik, but with the civil rights drama still unfolding in Little Rock: "The outstanding political fact of the present is that the Republicans have managed to recoup—through the integration issue—their sagging political fortunes." By taking up the cause of integration, "[the Republicans have] a potent weapon which chews the Democratic party to pieces so efficiently that it cannot be an effective opposition." The troops were still in Little Rock, and in upcoming battles segregationist southern Democrats would make matters worse. The party of Abraham Lincoln now had a leader in the White House willing and able to take a stand. Reedy concluded: "The only possibility is to find another issue which is even more potent. Otherwise the Democratic future is bleak." The other issue, the one that would save the party, would be Sputnik and America's eventual role in outer space. The memo ended by saying that if Sputnik gave partisan advantage to the party, it would also give an advantage to the American people. "This may be one of those moments in history when good politics and statesmanship are as close together as a hand in a glove." The Reedy memo became the battle plan that launched the Sputnik hearings, which began on November 25. The memo of October 17, 1957, is in box 420 of the LBJ Senate papers at the LBJ Library. There is an interview on the subject in NASA Monographs in Aerospace History *(Legislative Origins of the National Aeronautics and Space Act of 1958),* an oral history workshop held on April 3, 1992.

The Johnson hearings still make for compelling reading. The intensely political Johnson attempted to show that the Russians had staged a sneak attack, with the Pearl Harbor analogy coming up repeatedly. He seemed incredulous when he was told that the Russians had made periodic announcements of what they were up to. He drove home the point that the administration had done incalculable harm to the nation by permitting the Soviets to go into space first. "The Roman empire controlled the world because it could build roads. Later, when men moved to the sea, the British Empire was dominant because it had ships. Now the Communists have established a foothold in outer space," he said. "It is not very reassuring to be told that next year we will put a better satellite into the air. Perhaps it will even have chrome trim and automatic windshield wipers!"

Von Braun and Medaris were given hours to testify and carefully paint a picture of poor leadership in Washington. Addressing an issue that had been discussed in Washington for months, committee staffer Cyrus Vance, later secretary of state, asked Medaris, "General, is it correct that at one point when express orders were sent down to the Army not to launch a satellite, auditors from the Budget Bureau checked on your agency to make certain the orders were obeyed?" Major General Medaris replied: "I don't remember what bureau they came from. I know there were some people who came down to take a look and make sure that I was not fudging." Earlier in the month, in an interview with the *Huntsville Times,* Medaris had confirmed that the men sent to check on him were from the Bureau of the Budget. "Since I wasn't fired, I guess we got a 'clean bill.' "

Johnson's line of questioning was used again and again to show that Eisenhower and his national security advisers were doing all that they could to hamstring the Army, which if left to its own devices could have single-handedly beaten the Russians into space. Whether or not Eisenhower actually sent members of his Budget Bureau, an agency of the White House, to Huntsville to check on Medaris and von Braun is not as important as the fact that Huntsville wanted to play the role of a martyr held in check by Pentagon civilians and an old soldier in the White House, who they thought was closer to the Army that used mules than to one that used missiles.

Gavin dropped a bombshell during his December 13 testimony when he advocated the abolition of the Joint Chiefs of Staff as a way of managing the military. As a background to the hearings, Gavin announced that he would retire. Johnson replied, "This committee and its chairman are not going to tolerate any administration rubber-hose tactics or any effort to put committee witnesses in a straight-jacket." Now and again in the summer when his book came out, Gavin used his retirement as a moment of high drama. Medaris would stage a similar exit in 1960.

Behind the scenes, there was great concern over the first Vanguard launch attempt. Secretary of the Navy Thomas S. Gates sent a cautionary memo to Secretary of Defense McElroy on October 22: "I am gravely concerned about the effects of the Presidential statement of 9 October on the dates of intended firings of Vanguard vehicles. It said in effect that test vehicles were planned to be launched in December and a fully instrumented satellite in March 1958. The statement has been widely—though erroneously—interpreted in the press and within the administration as a commitment to achieve satellites on those dates. There is in fact only a probability—not a certainty—that satellites will be achieved on these first attempts." He ended the memo by noting that he was warning McElroy about the possibility of failure before the event rather than have to explain after. McElroy responded in a memo on October 29, which left no doubt about the commitment to get the satellites into orbit: "Your memorandum of 22 October expressed concern that the Presidential statement of 9 October on the U.S. satellite program has committed the Navy to meet the December 1957 launching of a test vehicle and a March 1958 launching of an instrumented satellite. Subsequent to the above, I have received a confirmation from the President that he expects the Department of Defense to meet these commitments."

The long Thanksgiving weekend began with a moment of reassurance as President Eisenhower, three days after his stroke, attended church. Billy Graham reacted privately and in writing: "Just a note to say that you are about the most remarkable man in history. I think your going to church last Thursday morning sent a sigh of relief throughout the entire world." He called the president "God's man of the hour" and commended him for being frank and candid with the American people in dealing with Russia's advances. Graham's answer to the Sputnik crisis was faith in God and attitude: "To win this battle I think it is going to take even more emphasis on the need for sacrifice, belt-tightening and renewed dedication, if the American way of life is going to be preserved. We Americans are growing soft. Our amusements and greed for money is acting as a sedative. We almost need laws for compulsory scientific training in all our schools, as well as compulsory physical training for all our young people. In other words, we need to toughen up!"

Ross Mark, who had covered the Vanguard launch for London's *Daily Express,* told me in 2000 that he watched as white smoke turned black and then "everything burst into flames." The smoke and fire confused Mark and the other reporters, who had never seen a liftoff before. Some actually thought that it had succeeded: "Four, three, two, one, zero," Harry Reasoner said live on CBS television at the cape. Then he said that it was a beautiful launch, and that he was amazed by its faster-than-the-eyes-could see ascent into space. "It was so quick," he naively told a nationwide audience, "I really didn't see it." Moments later, he realized that the reason he had not seen it shoot into space was that it had risen only a few inches. Reasoner, like the other reporters who had been invited to the launch to restore confidence in America, had no option but to report on an unmitigated failure. Columnist Dorothy Kilgallen asked, "Why doesn't somebody go out there, find it and kill it?"

The Johnson hearings themselves generated powerful and disturbing post-Vanguard headlines. On Saturday, December 14, for instance, von Braun testified that a vehicle the size of Sputnik II could be used to drop a hydrogen bomb on any U.S. city. This had been said before, but here it was coming from America's top rocket man in sworn testimony before a Senate committee. "Sputnik Called H-Threat" was the Sunday morning headline in the *Nashville Tennessean;* it included the subtitle "Von Braun Says It Can Bomb U.S."

The Sputniks had created a great opportunity for the Democrats in general and Lyndon Johnson in particular. He was making points that he would use in his bid for the presidency against Richard M. Nixon, the Republican candidate in waiting. In fact, the only member of the Eisenhower administration to sound at all alarmed by Sputnik was Vice President Nixon, who called it "a spectacular event" and went on to say that no event could have done as much to remind the United States and the USSR of the "increasingly terrifying aspects of modern warfare." But Nixon felt that he had to say something to prevent the Democrats from stealing all of his thunder—thunder he would need in the upcoming presidential election.

Green and Lomask *(Vanguard)* detail the aftermath of the explosion for the Vanguard team, including a morale-sapping fight that broke out between contractors and subcontractors as to what had caused the Vanguard explosion. The cause was finally determined to be "indeterminate" while changes were made. The Vanguard team, at this point, did not have the time or luxury for a debilitating family feud. The irony of Vanguard was that only people at the very top knew it was a scientific experiment that had had the role of America's technological savior foisted on it. Eisenhower understood this and sent the Vanguard team messages of support. Vice President Nixon wrote to the Vanguard team: "[At] a time when you have been 'catching it' from all sides, I want you to know that I, for one, feel you should have every support. . . . Keep up the good work." Eisenhower was confident that America would succeed soon enough and could not understand what all the fuss was all about. The president put a pertinent question to Detlev Bronk, president of the National Academy of Sciences: "Were we Americans the first to discover penicillin?" "You know the answer to that, Mr. President" was Bronk's reply. "And did we kill ourselves because we didn't?" Eisenhower asked. Bronk said that the president knew the answer to that, too.

The extent to which the failure was misunderstood was underscored a few weeks after the explosion, when Don Markarian, one of the Vanguard engineers, returned to his home in Baltimore, where he attempted to hire a housepainter. "Finally," he later recalled in Stehling (*Project Vanguard*), "one of the men I approached had the courtesy to level with me. 'To tell you the truth, Mr. Markarian, I don't feel much like working for anyone connected with Project Vanguard.'" Markarian added, "From the quantity of criticism that came hurling at us, you'd have thought we had committed treason."

With the Sputniks, the Soviets succeeded beyond all predictions, but some took solace in the fact that there seemed to be something crass, crude, and imperialistic about the "prodigal" use of power in sending up a satellite. Surmising correctly that the Russians had used three rocket clusters for each of the first two Sputniks, F. J. Krieger of the RAND Corporation, the Air Force think tank in Santa Monica, California, wrote in his 1958 report (*Behind the Sputniks*), "It is obvious that the Soviets in their struggle for world domination are applying their sledge-hammer technique not only to territorial affairs but also to the conquest of cosmic space."

Roberts (*First Rough Draft* and *The Nuclear Years*) discusses the Gaither report and his story for the *Washington Post*, which began with these ominous words: "The still top-secret Report portrays a United States in the gravest danger in its history. It pictures the nation moving in frightening course to the status of a second-class power. It finds America's long-term prospect one of cataclysmic peril in the face of rocketing Soviet military might and of a powerful, growing Soviet economy and technology which will bring new political, propaganda, and psychological assaults on freedom all around the globe. In short, the report strips away the complacency and lays bare the highly unpleasant realities in what is the first across the board survey of the relative postures of the United States and the Free World and of the Soviet Union and the Communist orbit." Roberts's account went on to mention the billions of dollars that would be required to build missiles and fallout shelters.

The NSC notes of January 16, 1958, available on-line from the Eisenhower Library, are most revealing on the point of the reaction by the two top men to fallout shelters. Nixon was even more blunt and to the point than Eisenhower, suggesting "that it be assumed that 40 million people would be killed in the event of all-out nuclear war. . . . If 40 million were killed, the United States would be finished."

The impact of the Sputniks on the UFO culture was huge. *The Flying Saucer Review's World Roundup of UFO Sightings and Events* contained this explanation on its dust jacket: "1957's big news story—the launching of the first earth satellite—focused the eyes of the world on the heavens, producing UFO sightings in ever-increasing numbers." For years to come, many ufologists maintained that the Sputniks, especially Sputnik II, were an attraction luring extraterrestrials. According to Trench (*The Flying Saucer Story*), "Sightings snowballed on a worldwide scale immediately after Sputnik II was launched." Hough and Randles (*The Complete Book of UFOs*) discuss the nights of November 2–3, 1957, when "UFO activity was enormous." This was either "an extraordinary coincidence, or . . . strong evidence that UFOs are being intelligently controlled and made a direct response to these shattering events. On the night that Laika became our first space traveler, fantastic things were afoot. The most astonishing set of close encounters struck the USA in what appears to be a blatant demonstration of superior technology." The authors detail thirty-six Laika-related cases in which vehicles experienced close encounters with glowing objects that impeded or destroyed their electrical systems.

EIGHT. AMERICAN BIRDS

Van Allen's IGY experiences are detailed in Hanle ("The Beeping Ball That Started a Dash into Outer Space") and Thomas (*Men of Space*). Stuhlinger had been trying to recruit Van Allen since 1952, when the original von Braun satellite proposal was for an inert object without a payload. If von Braun was to have a scientific package for "Slug," he would have to recruit a scientist. Von Braun had given Stuhlinger the job of finding an "honest to goodness" scientist for *his* satellite so that it would be accepted and even supported by the American scientific community. In November 1956, as Van Allen worked on the Vanguard instrumentation package in a University of Iowa basement, he was paid a visit by one of the most brilliant members of von Braun's Rocket Team. Stuhlinger told Van Allen that the Navy project was experiencing delays and pointed to information on the first test flight of the Jupiter C on September 20, during which it had traveled a record 3,100 miles and could have been put into orbit on that occasion.

The secrecy was imposed because of the drastic measures that had been put in place when Orbiter was canceled. The severity of these measures might have been exaggerated, but they were believed by those hiding their prohibited satellite work. According to Hibbs: "There was a story out of Huntsville that the Government Accounting Office went to ABMA and confiscated all reports which bore the word 'satellite' in their title and any equipment which looked like it had anything to do with satellites." Both this Hibbs quotation and the one in the text appear in Thomas (*Men of Space,* vol. 5, p. 72). The claim about the General Accounting Office is a hard one to pin down and, though believed at the time, may be mythical.

There are two bulging files of notes, clippings, and transcribed accounts of the Explorer launch in the Medaris papers at Florida Tech. John Hinrichs, a participant, put together a minute-by-minute set of notes on what was going on in the Pentagon on "Explorer" night. They have been relied on heavily in this narrative. The notes attest to the banter and spirit of the night. On waiting: "Mr. Brucker, enthusiastic and jovial, but a bit anxious too, made several good cracks. One 'Boys! This is just like waiting for the precincts to come in!' "

The White House reaction to Explorer was well documented by John Eisenhower *(Strictly Personal),* the president's son who was working there at the time. As he wrote in his memoir, he "sensed a feeling in the White House that the Army represented an upstart, second only to the Soviets."

Medaris was hardly shocked by the effigy burning. Photos of the display were carefully preserved in his personal archives in a special file.

Glennan *(The Birth of NASA)* and Bilstein *(Orders of Magnitude)* detail the transformation of NACA into NASA. The former had been created in 1915 as a national aeronautical advisory committee and aeronautical research laboratory. Over the years it had worked closely with the military while staying independent and under civilian control. Bilstein explained that NACA's "greatest political asset was its peaceful, research-oriented image. President Eisenhower and Senator Johnson and others in Congress were united in wanting above all to avoid projecting cold war tensions into the new arena of outer space."

Even before there was a NASA, a consensus was building in Washington to move many of the Army missile experts, including von Braun and his team, to the new civilian agency. But in *Newsweek*'s gossipy "Periscope" section for November 10, 1957, von Braun was quoted as saying that he would quit if Eisenhower put him under civilian control. Why? Because, said Braun, he wouldn't have the same "scientific freedom" under civilians that he had under the military.

The second Vanguard failure is described in Stehling *(Project Vanguard):* "A couple of us looked out of the tinted armored glass window into the inky sky. There we saw a heartbreaking sight. A long meteoric streak of flame ran like a luminous tear across the black visage of the sky and down through the clouds. We saw bits and pieces of shimmering metal illuminated by fire drifting down through the clouds, like infernal confetti. I heard a groaning in the blockhouse. Then silence. . . . Again there was a great croak in the news media. Not only had the Army beaten us, but we were not even able to follow." The fact that this was a test vehicle seemed to be forgotten, and rumors began circulating that the program might be canceled. The Vanguard managers tried to tell anyone who would listen that the Viking rocket test vehicles were not launch vehicles and that, as with any new system, early tests would fail. The only positive aspect to this debacle was that another disaster on the ground was averted and there was no humiliating film on television of an explosion and fireball, just an unphotogenic streak across the sky.

On March 7, 1958, still another launch attempt of a Vanguard satellite was scrubbed. A propellant leak was at fault, and the next day a safety problem, fog, and a technical malfunction conspired to hold it back again. The third attempt, on March 12, was halted four minutes before liftoff when pressurization problems occurred. Finally, on March 17, the countdown began for 7:00 A.M., but a hold was put on the launch because Explorer was passing overhead. Kurt Stehling would later recall this moment: "I must confess that never in my earlier life did I expect to see the day when one would have to wait until satellite traffic in the sky was cleared for the launching of another orbiter."

The Chivas Regal anecdote was retold by John Eisenhower *(Strictly Personal)*.

Although it is clear today that Vanguard was an experiment that succeeded through trial and error, it took some time before this sunk in. Unfortunately, few outside the world of space science understood Vanguard's value, perhaps because, in the words of *Saturday Review* science writer John Lear ("The Moon That Refused to Be Eclipsed"), the accomplishments "accumulated so slowly and undramatically during the last two years as to pass virtually unnoticed by the crowd."

The Van Cliburn "victory" was in keeping with the inclination of the press to treat Sputnik in terms of a contest that had been lost. The Sputnik launch had in fact come in the wake of a stunning athletic defeat. The 1956 Summer Olympic Games, held in Melbourne, Australia, were the first held in the Southern Hemisphere, where summer came late, and were almost called off because of turmoil around the world. The British and the French had invaded Suez after the Egyptian seizure of the canal, and on the eve of the competition, the Russians used tanks and troops to brutally suppress an attempted uprising in Hungary aimed at overthrowing the Soviet-dominated regime. Sympathy for Hungary was intense in the United States, and the American team was intent on avenging the honor of the courageous Hungarian "freedom fighters" by beating Russia. It failed. American competitors picked up thirty-two gold medals, but Russia emerged as the top nation with a total haul of ninety-eight medals, thirty-seven of them gold. The heroics at the Olympics were performed by members of the Hungarian water polo team; they took their war with the Russians into the pool, which became a literal bloodbath. Police were called in to calm the two teams and spectators alike after a Hungarian player's eyebrow had been gashed during a clash with a Russian opponent. The Hungarians went on to win the match 4–0 and the gold medal.

As for the myth that the Explorer was created from scratch: A 1997 "Data Sheet" issued by the Smithsonian's National Air and Space Museum on "The First United States Satellite and Space Launch Vehicle" reports, "Working closely together, ABMA and JPL completed the job of modifying the Jupiter-C and building the Explorer-I in 84 days." What tended to be forgotten at the time was that Explorer went aloft on a glorified version of the V-2 built by a team that had been together for a quarter of a century.

The other thing seldom mentioned was that the opposite happened to the Vanguard project. When the project began it was assumed that it would use the established Viking engineering team at Martin; but by the time Vanguard was assigned to Martin, the team had been broken up, with most of its top engineers going to high-priority missile projects.

The story of the Lunik kidnapping was released in 1994 from the report originally appearing in the winter 1967 issue of *Studies in Intelligence.* Oddly, even after the report had been declassified, the CIA denied any such involvement. My copy of the report comes from Charles Vick of the Federation of American Scientists, not the CIA. Park's report appears in *Voodoo Science* and is quoted here with his permission. Park posts a provocative weekly electronic column on science issues at http://www.aps.org/WN/

NINE. IKE SCORES

Glennan *(The Birth of NASA),* Bilstein *(Orders of Magnitude),* and the chronological files at the NASA History Office in Washington, D.C., provided details on the formation of NASA. The Medaris anti-NASA speeches are on file at the Medaris archives. In March in San Francisco he declared to members of the Commonwealth Club that it would take too much time to create a civilian space agency and that this would put the Russians farther ahead. "No organization except the military has the experience," adding with irritation, "I've never understood why a man should be considered incompetent because he wears a uniform." The Goodpaster quote is from an interview with Eugene M. Emme and A. F. Roland of July 22, 1974, on file at the NASA History Office.

The prime sources on SCORE include Chapman *(Atlas),* Dwight D. Eisenhower *(The White House Years),* Davis ("The Talking Satellite"), and Wainwright ("How Insiders Kept Their Great Secret"). A SCORE file at the NASA History Office makes important links between SCORE and earlier efforts. The U.S. Navy had begun communications experiments bouncing radio signals off the Moon in 1954. The world's first operational space communications system, called Communication by Moon Relay (CMR), was used between 1959 and 1963 to link Hawaii and Washington, D.C. The application to communications of a spacecraft that remains stationary with respect to the surface of the Earth was invented by Sir Arthur C. Clarke and presented in his article "Extra-Terrestrial Relays" (in *Wireless World,* October 1945, pp. 305–8).

O'Keefe *(Nuclear Hostages)* is an important source on Teak and Orange. The NASA history office has a copy of von Braun's Hawaii speech. (In a Strangelovean appeal to local pride, he crowed, "It should not surprise the residents of Hawaii that the two shots were by far the most spectacular ever fired by this country and—it may be presumed—by any country.") A one-page, undated document titled "Nuclear Explosion over the Pacific Ocean" was located in the file on "Teak/Orange" in the History office at NASA headquarters in Washington, D.C.

CORONA's accomplishments are covered in Day, Logsdon, and Latell *(Eye in the Sky),* Ruffner *(Corona),* and Peebles *(The Corona Project).* Recently released National Intelligence Estimates reveal that CORONA got a lot of work done besides debunking the missile gap that had been blamed on Eisenhower and Nixon. Among CORONA's accomplishments are the following (taken from an official list): detected all Soviet medium-range, intermediate-range, and ICBM complexes; imaged each Soviet submarine class from deployment to operational bases; provided inventories of Soviet bombers and fighters; revealed the presence of Soviet missiles in Egypt, protecting the Suez Canal; identified Soviet nuclear assistance to the People's Republic of China;

monitored the SALT I treaty; uncovered the Soviet ABM program and sites (GALOSH, HEN HOUSE, etc.); identified Soviet atomic weapon storage installations; identified People's Republic of China missile launching sites; determined precise locations of Soviet air defense missile batteries; observed construction and deployment of the Soviet ocean surface fleet; identified Soviet command and control installations and networks; provided mapping for Strategic Air Command targeting and bomber routes; identified the Plesetsk Missile Test Range, north of Moscow.

The witness to Kennedy's hope that the lunar landing would occur during his second term was Paul Means, then serving on the National Space Council. Means, who also worked for *Missiles and Rockets* during the early days of the space race, vetted an early version of this manuscript for accuracy.

Several other indictments were made against the Sputnik-driven timetable at the time of the Apollo fire. Betty Grissom, Virgil (Gus) Grissom's widow, wrote in *Starfall,* her personal memoir of the race for the Moon, that the work on space had been proceeding at a normal pace until the Sputnik launch; but then scientists and engineers were abruptly forced to compress long-range engineering decisions into immediate ones that would deliver functioning equipment "yesterday if not sooner."

For many years, only a few insiders knew that the United States had taken its first quiet step toward the lunar surface in the spring of 1957, half a year before Sputnik and four years before President Kennedy declared the exploration of the Moon as a national goal. At that moment the Rocket Team began designing the Saturn booster to be ten times more powerful than their own powerful Jupiter. This tenfold increase in thrust would propel a space probe not only out of Earth's orbit but presumably out of the hands of those in Washington who would limit the Army's space dreams. It was during the previous November that Secretary of Defense Charles E. Wilson had assigned responsibility for all intermediate- and long-range missiles to the Air Force. If the Army was to stay in the big-rocket business, it would have to find work for itself, and a superrocket capable of space exploration became the Army's ticket into space. What Gavin, Medaris, von Braun, and those under them had not planned on was the national urge to make space a civilian realm. The eventual lunar landing was as much a validation of the value of the Medaris–von Braun partnership as Explorer.

Medaris, later an Episcopal priest and then a priest in the even more conservative Anglican-Catholic church, let bitterness get in the way of enjoying his historic role in space. The tough-talking and determined two-star general uttered unbecoming lines to reporters on NASA ("It stinks") and the fact that he did not receive more attention ("I was a household word, and now I go over to Cape Canaveral and I have to spell my name," he complained in a 1978 interview). He passed away in 1990. His papers are an invaluable resource for understanding the early days of the U.S. space program.

In 1997 Apollo 17 astronaut Harrison "Jack" Schmitt revealed at a Washington, D.C., news conference that Soviet émigrés had told him that the display of rocket power on Apollo 8 caused the Russians to give up any hope of beating America to the Moon. For all intents and purposes, the race was over.

TEN. SPUTNIK'S LEGACY

Medaris and Heatter were not the only ones who credited Sputnik with spurring America on to greatness in celestial as well as terrestrial realms—first, as a spacefaring nation, then as a technological superpower. A few others saw it coming and offered thanksgiving. "We needed Sputnik. It is sure proof that God has not despaired of us," said former governor and presidential candidate Adlai Stevenson, while Wernher von Braun commented, "The Russians have given us Americans a free lecture. We better put it to good use."

Stephen King's account appears in his *Danse Macabre,* a nonfiction work that provides a guided tour of America's relationship with fear. Doug Evelyn's high school paper on Sputnik resides in a file in his basement.

The recalling of Sputnik anniversaries was aided greatly by still another resource in the NASA History Office: a series of thick files labeled "Sputnik—First Anniversary," "Sputnik—Fifth Anniversary," and so on. The dispatch from Moscow on the thirty-fifth anniversary of Sputnik was abstracted from a Reuters report, which had equivalents—also festooned with irony—from other sources. Here is an AP dispatch:

> MOSCOW—*The fate of the Exhibition of Economic Achievements of the U.S.S.R. where the triumphs of Soviet man were celebrated as nowhere else in dazzling exhibits encased in 80 World's Fair–style pavilions is stunning. The grandest of the pavilions, once filled with the Sputnik replicas and space paraphernalia, has been cleaned out to make room for an enormous Western car showroom. Colleen Jordan of the Associated Press visited the hall in 1993 and followed Valery Frantsev, 47, who had brought his 4-year-old son, Slava, here especially to see the space exhibit that had made him gasp with excitement in his own childhood. "To his shock," said the AP dispatch, "Mr. Frantsev found the giant photograph of the space hero Yuri Gagarin beaming out upon rows of Cadillacs, Dodge Caravans and Jeep Grand Cherokees." "We had many achievements," Mr. Frantsev said. "Where are they?" But his son was completely transfixed. He took a quick look at the remaining Sputnik, then carefully absorbed every detail of every car. "Mercedes Benz," he said softly at one point. "This is what makes his eyes light up," Mr. Frantsev said, gesturing toward a cherry-red, $45,000 Cadillac de Ville.*

One of the more interesting items in these files is a story in which the claim was made that the world was still unable to accept the enormity of the event. "We're down to 40 years later and people don't understand the effect of Sputnik. Can you imagine what's going on when you layer new technology upon technology? All the effects that are happening on us subliminally?" said Nelson Thall, president of the Marshall McLuhan Center, a Toronto-based group dedicated to broadening the research of the 1960s media guru, in a December 30, 1994, Associated Press interview. "The accumulation of new devices and habits can be daunting," Thall added. In his 1964 *Understanding Media* (New York: McGraw-Hill, 1964), McLuhan had called the first Sputnik "a witty taunting of the capitalist world by means of a new kind of technological image or icon, for which a group of children in orbit might yet be a telling retort."

APPENDIX: SPUTNIK'S LONG, LEXICAL ORBIT

This word history was prepared with the aid of the citation system at Merriam-Webster in Springfield, Massachusetts, and Dave Kelly of the Library of Congress, who helped me find the elusive work of Smal-Stocki *(The Impact of the "Sputnik" on the English Language of the U.S.A.).* Dalzell *(Flappers 2 Rappers)* was essential to documenting the Sputnik / Beatnik relationship.

Bibliography

PRINT SOURCES

Adams, Sherman. *Firsthand Report: The Story of the Eisenhower Administration.* New York, Harper and Brothers, 1961.

Akens, David S. *A Picture History of Rockets and Rocketry.* Huntsville, Ala.: Strode, 1964.

Alexander, Charles C. *Holding the Line: The Eisenhower Era, 1952–1961.* Bloomington: Indiana University Press, 1975.

Allen, William H., ed. *Dictionary of Technical Terms for Aerospace Use.* Washington, D.C.: National Aeronautics and Space Administration, 1965. (NASA SP-7)

Alston, Giles. "Eisenhower: Leadership in Space Policy." In *Reexamining the Eisenhower Presidency,* ed. Shirley Anne Warshaw. Westport, Conn.: Greenwood Press, 1993.

Ambrose, Stephen E. *Eisenhower.* New York: Simon and Schuster, 1984.

———. *Ike's Spies: Eisenhower and the Espionage Establishment.* Garden City, N.Y.: Doubleday, 1981.

Anderson, Frank W., Jr. *Orders of Magnitude: A History of NACA and NASA, 1915–1980.* Washington, D.C.: National Aeronautics and Space Administration, 1981. (NASA SP-4403)

Angle, Paul M. *The American Reader.* New York: Rand McNally, 1958.

Asimov, Isaac. *Exploring the Earth and the Cosmos.* New York: Crown, 1982.

———. *In Joy Still Felt: The Autobiography of Isaac Asimov, 1954–1978.* Garden City, N.Y.: Doubleday, 1980.

Bainbridge, William Sims. *The Spaceflight/Revolution.* Malabar, Fla.: Krieger, 1983.

Baker, David. *The History of Manned Spaceflight.* New York: Crown, 1981.

———. *The Rocket.* London: New Cavendish Books, 1978.

Baker, Michael E. *Redstone Arsenal: Yesterday and Today.* Huntsville, Ala.: U.S. Army Missile Command Redstone Arsenal, 1988.

Barrett, George. "Visit with a Prophet of the Space Age." *New York Times,* October 20, 1957.

Barton, William S. "First Attempt to Create a Moon Failed at New Mexico Nine Years Ago." *Los Angeles Times,* July 30, 1955.

Beahm, George. *The Stephen King Story.* Kansas City: Andrews and McMeel, 1991.

Benedict, Howard. "Dr. Van Allen Decries Space Budget Cuts." *Huntsville Times*, January 28, 1968.
Berg, A. Scott. *Lindbergh*. New York: Putnam, 1998.
Bergaust, Erik. *Murder on Pad 34*. New York: Putnam, 1968.
———. *Wernher von Braun*. Washington, D.C.: National Space Institute, 1976.
Bergaust, Erik, and William S. Beller. *Satellite! The First Step into the Last Frontier*. Garden City, N.Y.: Doubleday, 1958.
Bergaust, Erik, and Seabrook Hull. *Rocket to the Moon*. New York: Von Nostrand, 1958.
Berkner, Lloyd V. "Earth Satellites and Foreign Policy." *Foreign Affairs*, January 1958.
———. "Man's Space Satellites." *Bulletin of the Atomic Scientists*, March 1958.
Beschloss, Michael R. *The Crisis Years: Kennedy and Khrushchev, 1960–1963*. New York: Harper, 1991.
———. *Mayday: The U-2 Affair*. New York: Harper Perennial, 1986.
Biddle, Wayne. "Science, Morality, and the V-2." *New York Times*, October 2, 1992.
Bilstein, Roger E. *Orders of Magnitude: A History of the NACA and NASA, 1915–1900*. Washington, D.C.: Government Printing Office, 1989. (NASA SP-4206)
———. *Stages to Saturn: A Technological History of the Apollo/Saturn Launch*. Washington, D.C.: National Aeronautics and Space Administration, 1989. (NASA SP-4206)
Binder, L. James. "The Night the Army Boosted America into the Space Age." *Army*, February 1965.
Bird, Kai. *The Color of Truth*. New York: Simon and Schuster, 1998.
Bissell, Richard M., Jr., "Origins of the U-2." *Air Power History*, fall 1989.
Bissell, Richard M., Jr., with Jonathan E. Lewis and Frances T. Pudlo. *Reflections of a Cold Warrior: From Yalta to the Bay of Pigs*. New Haven, Conn.: Yale University Press, 1996.
Black, Harold D. "Early Development of Transit, the Navy Navigation Satellite System." *Journal of Guidance, Control, and Dynamics*, July–August 1990.
Bloomfield, Lincoln, ed. *Outer Space: Prospects for Man and Society*. Englewood Cliffs, N.J.: Prentice-Hall, 1962.
Boffa, Giuseppe. *Inside the Khrushchev Era*. New York: Marzani and Munsell, 1959.
Boffey, Philip M., William J. Broad, Leslie H. Gelb, Charles Mohr, and Holcomb B. Noble. *Claiming the Heavens: The New York Times Complete Guide to the Star Wars Debate*. New York: Times Books, 1988.
Boorstin, Daniel J. *The Americans: The Democratic Experience*. New York: Random House, 1973.
Booth, T. Michael, and Duncan Spencer. *Paratrooper: The Life of Gen. James M. Gavin*. New York: Simon and Schuster, 1994.
Bower, Tom. *The Paperclip Conspiracy*. Boston: Little, Brown, 1986.
Bowler, Mike. "Is It Time for a New Jolt to Education?" *Baltimore Sun*, October 2, 1982.
Bowles, Chester. "The New Challenge." In *The Britannica Book of the Year*. Chicago: Encyclopedia Britannica, 1958.
Bradley, Wendell. " 'Sputnik' Hunt Recalled by FCC Award Winner." *Washington Post*, April 10, 1959.
Braun, Wernher von. *The Mars Project*. Champaign: University of Illinois Press, 1991. (Reprint of von Braun's original 1952 designs for manned landings on Mars.)
———. *Space Frontier*. New York: Fawcett Books, 1969.
———. "Where Are We in Space?" *This Week*, January 1, 1961.

Braun, Wernher von, and Frederic I. Ordway, III. *History of Rocketry and Space Travel.* New York: Crowell, 1969.

———. *The Rockets' Red Glare.* Garden City, N.Y.: Anchor Press/Doubleday, 1976.

Brendon, Piers. *Ike: His Life and Times.* New York: Harper and Row, 1986.

Breuer, William B. *Race to the Moon: America's Duel with the Soviets.* Westport, Conn.: Praeger, 1993.

Bright, Charles D., ed. *Historical Dictionary of the U.S. Air Force.* Westport, Conn.: Greenwood Press, 1992.

Broad, William J. "U.S. Planned Nuclear Blast on the Moon, Physicist Says." *New York Times,* May 16, 2000.

Brooks, Courtney G., Hames M. Grimwood, and Lloyd S. Swenson. *Chariots for Apollo.* Washington, D.C.: National Aeronautics and Space Administration, 1989. (NASA SP-4205)

Brooks, John. *The Great Leap: The Past Twenty-five Years in America.* New York: Harper and Row, 1966.

Buedeler, Werner. *Operation Vanguard.* London: Burke, 1958.

Bulkeley, Rip. "Harbingers of Sputnik: The Amateur Radio Preparations in the Soviet Union." *History and Technology* 16 (1999): 67–102.

———. *The Sputniks Crisis and Early United States Space Policy: A Critique of the Historiography of Space.* Bloomington and Indianapolis: Indiana University Press, 1991.

Burgess, Colin. "Celebrating Sputnik—Forty Years On." *Spaceflight,* October 1997.

Burrough, Bryan. *Dragonfly: NASA and the Crisis Aboard Mir.* New York: HarperCollins, 1998.

Burrows, William E. *Deep Black.* New York: Random House, 1986.

———. *This New Ocean.* New York: Random House, 1998.

Bush, Vannevar. *Modern Arms and Free Men.* New York: Simon and Schuster, 1949.

———. *Pieces of the Action.* New York: William Morrow, 1970.

Caidin, Martin. *Overture to Space.* New York: Duell, Sloan and Pearce, 1963.

———. *Red Star in Space.* New York: Crowell-Collier, 1963.

———. *Rendezvous in Space.* New York: Dutton, 1962.

———. *Vanguard.* New York: Dutton, 1957.

Campbell, Margaret A. "America's Reaction to Sputnik." (Unpublished thesis produced at Coastal Carolina College and on file at the NASA history office in Washington, D.C.)

Canan, James W. *The Superwarriors.* New York: Weybright and Talley, 1975.

———. *War in Space.* New York: Harper and Row, 1982.

Canby, Courtland. *A History of Rockets and Space.* New York: Hawthorne Books, 1963.

Canfield, Cass. *Up and Down and Around.* New York: Harper's Magazine Press, 1971.

Canizares, Alex. "Sputnik Was More Advanced Than U.S. Admitted, Historian Uncovers." space.com, November 1, 1999.

Carreau, Mark. "With Sputnik Launch Thirty Years Ago, Space Race Was in Orbit." *Houston Chronicle,* October 4, 1987.

Carter, Hodding. "The Rocket Scientists Settle Down." *Collier's,* November 12, 1954.

Carter, Paul A. *Another Part of the Fifties.* New York: Columbia University Press, 1983.

Chaikin, Andrew. *A Man on the Moon.* New York: Viking, 1994.

Chang, Iris. *Thread of the Silkworm.* New York: Basic Books, 1995.

Chapman, John L. *Atlas: The Story of a Missile.* New York: Harper and Brothers, 1960.

Chapman, Sydney. "Mass Attack on the Earth's Mysteries." *Popular Mechanics,* November 1955.

Chartrand, Mark. "The Rocket Pioneers." *Ad Astra,* March 1989.
Childers, Frank. "First U.S. Satellite Launched Forty Years Ago." *Florida Today* (Melbourne), January 31, 1998.
Childs, Marquis. *Eisenhower: Captive Hero.* New York: Harcourt, Brace, 1958.
Clarke, Arthur C. *Astounding Days: A Science Fiction Autobiography.* New York: Bantam Books, 1989.
———. *The Exploration of Space.* New York: Pocket Books (Cardinal Edition), 1954.
———. *The Making of a Moon.* New York: Harper and Brothers, 1957.
———. *The Promise of Space.* New York: Harper and Row, 1968.
———. *The View from Serendip.* New York: Ballantine Books, 1978.
———, ed. *The Coming of the Space Age: Famous Accounts of Man's Probing of the Universe.* New York, Meredith Press, 1967.
Clayton, Donald D. *The Dark Night Sky.* New York: Quadrangle (New York Times Book Company), 1978.
Cliburn, Van. "I Like My Success." *Parade,* May 24, 1959.
Cline, Ray S. *Secrets, Spies, and Scholars.* Washington, D.C.: Acropolis Books, 1976.
Cohn, Victor. "The Big Push in Soviet Science." *Saturday Evening Post,* January 31, 1959.
Colby, William. *Honorable Men: My Life in the CIA.* New York: Simon and Schuster, 1978.
Considine, Douglas M., ed. *Van Nostrand's Scientific Encyclopedia.* New York, 1968.
Collins, Martin J., and Sylvia D. Fries, eds. *A Spacefaring Nation. Perspectives on American Space History and Policy.* Washington, D.C.: Smithsonian Institution Press, 1991.
Collins, Michael. *Carrying the Fire: An Astronaut's Journeys.* New York: Farrar, Straus and Giroux, 1974.
———. *Mission to Mars.* New York: Grove Weidenfeld, 1990.
Cook, Blanche Wiesen. *The Declassified Eisenhower.* New York: Penguin Books, 1981.
Corliss, William R. "The Evolution of the Satellite Tracking and Data Acquisition Network (STADAN)." Goddard Historical Note No. 3. Greenbelt, Md., n.d.
Cowen, Robert C. "Declassified Papers Show U.S. Won Sputnik Race After All." *Christian Science Monitor,* October 23, 1996.
———. "The Scoop on Sputnik; A Science Writer's Reflections." *Christian Science Monitor,* October 1, 1997.
———. "What Eisenhower Did Not Reveal." *Christian Science Monitor,* October 1, 1997.
Cox, Donald W. *The Space Race: From Sputnik to Apollo . . . and Beyond.* Philadelphia: Chilton, 1962.
Cox, Donald W., and Michael Stoiko. *Spacepower—What It Means to You.* Philadelphia: John C. Winston, 1958.
Crakshaw, Edward. *Khrushchev: A Career.* New York: Viking Press, 1966.
Cronkite, Walter. *A Reporter's Life.* New York: Knopf, 1996.
Crouch, Tom D. *Aiming for the Stars: The Dreamers and Doers of the Space Age.* Washington, D.C.: Smithsonian Institution Press, 1999.
Dalzell, Tom. *Flappers 2 Rappers.* Springfield, Mass.: Merriam-Webster, 1996.
Daniels, Lynne. "Statement of Prominent Americans at the Beginning of the Space Age." NASA Historical Note No. 22. Washington, D.C.: NASA, 1965.
Daniloff, Nicholas. *The Kremlin and the Cosmos.* New York: Knopf, 1972.
Daso, Dik. "Origins of Airpower: Hap Arnold's Command Years and Aviation Technology, 1936–1945." *Air Chronicles,* winter 1996.

Davies, Merton E., and William R. Harris. *RAND's Role in the Evolution of Balloon and Satellite Observation Systems and Related U.S. Space Technology.* Santa Monica, Calif.: RAND Corporation, 1988. (R-3692-RC)

Davis, Deane. "The Talking Satellite: A Reminiscence of Project SCORE." *Journal of the British Interplanetary Society* 52 (1999).

Davis, Kenneth S. "Father of Rocketry." *New York Times,* October 23, 1960.

Day, Dwayne A. "Exhibiting the Space Race." *Quest: The History of Spaceflight Quarterly,* spring 1999.

———. "A Failed Phoenix: The KH-6 Lanyard Reconnaissance Satellite." *Spaceflight,* May 1997.

———. "Lifting the Veil." *Spaceflight,* August 1995.

———. "New Revelations About the American Satellite Programme Before Sputnik." *Spaceflight,* November 1994.

———. "Out of the Shadows: The Shuttle's Secret Payloads." *Spaceflight,* February 1999.

———. "A Strategy for Space: Donald Quarles, the CIA, and the Scientific Satellite Programme." *Spaceflight,* September 1996.

———. "Those Magnificent Spooks and Their Spying Machines." *Spaceflight,* March 1997.

———. "The Von Braun Paradigm." *Space Times: Magazine of the American Astronautical Society,* November–December 1994.

Day, Dwayne A., John M. Logsdon, and Brian Latell. *Eye in the Sky: The Story of the Corona Spy Satellite.* Smithsonian History of Aviation Series. Washington, D.C.: Smithsonian Institution Press, 1998.

de Ropp, Robert S. *The New Prometheans.* New York: Delacorte, 1972.

Dennis, Landt. " 'Loony Moon Man' Lands in New Mexico." *Christian Science Monitor,* January 30, 1971.

Dethloff, Henry C. *Suddenly, Tomorrow Came . . . A History of the Johnson Space Center.* Houston: National Aeronautics and Space Administration, Lyndon B. Johnson Space Center, 1993. (NASA SP-4307)

Diamond, Edwin. "Nick Christofilos: The Not-So Crazy Greek." *Saga,* October 1959.

———. *The Rise and Fall of the Space Age.* Garden City, N.Y.: Doubleday, 1964.

———. "That Old 'Missile Gap' Madness." *Newsweek,* October 11, 1982.

Diamond, Edwin, and Stephen Bates. "Sputnik." *American Heritage,* October 1997.

Divine, Robert A. *The Sputnik Challenge.* New York and Oxford: Oxford University Press, 1993.

Dorsey, Gary. "Sputnik Plus Forty Years." *Baltimore Sun,* October 5, 1997.

Douglass, John A. "A Sputnik Reflection." *Space Times,* January–February 1998.

Dow, Peter B. *Schoolhouse Politics: Lessons from the Sputnik Era.* Cambridge: Harvard University Press, 1991.

Dryden, Hugh L. "The IGY Man's Most Ambitious Study of His Environment." *National Geographic,* February 1956.

Duff, Brian. "Tovarich." *Air & Space,* December 1994.

Dulles, Allen W. *The Craft of Intelligence.* New York: Signet Books, 1965.

Dunn, Marcia. "American, Russian Rocket Debuts." Associated Press Wire, May 24, 2000.

Durant, Frederick C. *Between Sputnik and the Shuttle: New Perspectives on American Astronautics, 1957–1980.* AAS History Series, vol. 3. San Diego: Univelt, 1981.

Dusheck, George. "His Math Was Fuzzy, But Christofilos Had Right Idea on Space Explosions." *Washington Daily News,* March 20, 1959.

Dye, Lee. "Sputnik: Soviet Feat Brought Global Change." *Los Angeles Times,* October 3, 1987.

Eisenhower, Dwight D. *The White House Years: Waging Peace, 1956–1961.* Garden City, N.Y.: Doubleday, 1965.

Eisenhower, John S. D. *Strictly Personal.* New York: Doubleday, 1974.

Eisenhower, Susan. *Breaking Free: A Memoir of Love and Revolution.* New York: Farrar, Strauss and Giroux, 1995.

Emme, Eugene M. *The History of Rocket Technology.* Detroit: Wayne State University Press, 1964.

———. *A History of Space Flight.* New York: Holt, Rinehart and Winston, 1965.

———. *The Impact of Air Power: National Security and World Power.* Princeton: Van Nostrand, 1950.

———, ed. *Two Hundred Years of Flight in America: A Bicentennial Survey.* San Diego: American Astronautical Society, 1977.

Evans, Bergen. "New World, New Words." *New York Times Magazine,* April 9, 1961.

Ewald, William Bragg, Jr. *Eisenhower the President: Crucial Days, 1951–1960.* Englewood Cliffs, N.J.: Prentice-Hall, 1981.

Ezell, Edward Clinton, and Linda Neuman Ezell. *The Partnership: A History of the Apollo-Soyuz Test Project.* Washington, D.C.: National Aeronautics and Space Administration, 1978. (NASA SP-4209)

Fairman, Milton. "The Race to Explore." *Popular Mechanics,* March 1930.

Fanning, Leonard M. *The Man Behind the Rocket.* Detroit: General Motors Information Rack, 1966.

Feldbaum, Carl B., and Ronald J. Bee. *Looking the Tiger in the Eye: Confronting the Nuclear Threat.* New York: Vintage, 1990.

Ferguson, Eugene S. *Engineering and the Mind's Eye.* Cambridge: MIT Press, 1993.

Fields, Sidney. "First Sputnik, Then the World: Will We Catch Up to Russia? *New York Mirror,* November 24, 1957.

Flanagan, Mike. "Thirty Years Ago, the Future Was Launched." *Chicago Tribune,* October 4, 1987.

Flying Saucer Review. The Flying Saucer Review's World Roundup of UFO Sightings and Events. New York: Citadel Press, 1958.

Foley, John E. "A Teen and a Nation Are Spurred On by Sputnik." *Christian Science Monitor,* October 1, 1997.

Fontaine, André. *History of the Cold War.* New York: Pantheon Books, 1969.

Ford, Daniel. *The Button.* New York: Simon and Schuster, 1985.

Fortune, editors of. *The Space Industry: America's Newest Giant.* Englewood Cliffs, N.J.: Prentice-Hall, 1962.

Frankel, Max. "Soviet Scientists Insist Rocket Pieces Fell on Alaska." *New York Times,* December 8, 1957.

French, Bevan M., and Stephen P. Maran. *A Meeting with the Universe.* Washington, D.C.: National Aeronautics and Space Administration, 1981. (NASA EP-177)

Froehlich, Jack E., and Albert R. Hibbs. *Contributions of the Explorer to Space Technology.* Pasadena, Calif.: Jet Propulsion Laboratory, 1958.

Fuller, Helen. *Year of Trial: Kennedy's Crucial Decisions,* New York: Harcourt Brace, 1962.

Furgurson, Ernest B. "Father of Our Rocket." *Baltimore Sun,* November 24, 1957.

Furnas, Clifford C. "Birthpangs of the First Satellite." *Research Trends* (Cornell Aeronautical Laboratory), spring 1970.

———. "Why Did We Lose the Race? Critics Speak Up." *Life,* October 21, 1957.
———. "Why Vanguard?" *Life,* October 2, 1957.
Gall, Sarah L., and Pramberger, Joseph T. *NASA Spinoffs: Thirty Year Commemorative Edition.* Washington, D.C.: National Aeronautics and Space Administration, 1992.
Garber, Stephen J., compiler. *Research in NASA History: A Guide to the NASA History Program.* Washington, D.C.: National Aeronautics and Space Administration, 1997. (NASA HHR-64)
Gatland, Kenneth. *Illustrated History of Space Technology.* New York: Harmony Books, 1981.
Gatland, Kenneth, ed., with Wernher von Braun, Harry Ross, and A. V. Cleaver. *Project Satellite.* New York: British Book Centre, 1958.
Gavaghan, Helen. *Something New under the Sun.* New York: Copernicus, 1998.
Gavin, James M. *War and Peace in the Space Age.* New York: Harper and Brothers, 1958.
Gaynor, Frank. *Aerospace Dictionary.* New York: Philosophical Library, 1960.
Gemarekian, Edward. "Generals Back 'Satellite Eye." *Washington Post,* April 18, 1958.
Gibney, Frank, and George J. Feldman. *The Reluctant Space-Farers.* New York: New American Library, 1968.
Gilbert, Martin. *A History of the Twentieth Century.* Vol. 3, *1952–1999.* New York: William Morrow, 1999.
Gilchrist, Jim. "Spaced Out with Celestial Wonder." *Scotsman,* October 2, 1997.
Glasstone, Samuel. *Sourcebook on the Space Sciences.* Princeton, N.J.: D. Van Nostrand, 1965.
Glenn, John, with Nick Taylor. *John Glenn: A Memoir.* New York: Bantam Books, 1999.
Glennan, T. Keith. *The Birth of NASA: The Diary of T. Keith Glennan.* Ed. J. D. Hunley. Washington, D.C.: National Aeronautics and Space Administration, 1993. (NASA SP-4105)
Golden, William T. *Science and Technology Advice to the President, Congress, and the Judiciary.* Washington, D.C.: American Association for the Advancement of Science, 1993.
Goldstein, Stanley H. *Reaching for the Stars.* New York: Praeger, 1987.
Golovine, M. N. *Conflict in Space.* New York: St. Martin's, 1962.
Goodwin, Doris Kearns. *The Fitzgeralds and the Kennedys: An American Saga.* New York: Simon and Schuster, 1987.
———. *Wait Till Next Year.* New York: Simon and Schuster, 1997.
Goodwin, Richard N. *Remembering America.* New York: Little, Brown, 1988.
Gor'kov, V., and Avdeev Yu. *An A–Z of Cosmonautics.* Moscow: Mir Publishers, 1989.
Gould, John. "What Sputnik and Police Cruisers Set Off." *Christian Science Monitor,* May 13, 1997.
Green, Constance McLaughlin, and Milton Lomask. *Vanguard: A History.* Washington, D.C.: Smithsonian Institution Press, 1971.
Greenstein, Fred I. *The Hidden-Hand Presidency: Eisenhower as Leader.* New York: Basic Books, 1982.
Grissom, Betty, and Henry Still. *Starfall.* New York: Crowell, 1974.
Guggenheim, Harry F. "Goddard Deserves Niche as Father of Rocketry." *Evening Star,* July 16, 1969.
Gunston, Bill. *Rockets and Missiles.* London: Salamander, 1979.
Haber, Heinz. "Space Satellites, Tools of Earth Research." *National Geographic,* April 1956.

Hagen, John P. "The Viking and the Vanguard." Paper delivered to the program on the "History of Rocket Technology," Philadelphia, December 28, 1962.

Halberstam, David. *The Fifties*. New York: Villard Books, 1993.

Hall, R. Cargill. "Early U.S. Satellite Proposals." *Technology and Culture*, fall 1963.

———. "The Eisenhower Administration and the Cold War." *Prologue*, spring 1995.

———. "The Origins of U.S. Space Policy." *Colloquy*, December 1993.

———. "Vanguard and Orbiter." *Airpower Historian*, 9, no. 4 (October 1964).

Hall, R. Cargill, and Jacob Neufeld. *The U.S. Air Force in Space: 1945 to the Twenty-first Century*. Washington, D.C.: USAF History and Museums Program, 1998.

Hamby, Alonzo L. *The Imperial Years: The U.S. Since 1939*. New York: Weybright and Talley, 1976.

Hanle, Paul A. "The Beeping Ball That Started a Dash into Outer Space." *Smithsonian*, October 1982.

Hansen, James R. "If Sputnik Had Not Been First: Ideas for a Counter-Factual History of the U.S. Space Program." In *Inner Space, Outer Space: Humanities, Technology, and the Postmodern World*, ed. Daniel Schenker, Craig Hanks, and Susan Kray, pp. 161–68. Huntsville, Ala.: Southern Humanities Conference, 1993.

Harford, James. *Korolev: How One Man Masterminded the Soviet Drive to Beat America to the Moon*. New York: Wiley, 1997.

Harper, Harry. *Dawn of the Space Age*. London: Samson, Low, Marston, n.d.

Harris, Gordon L. *A New Command: The Life of Bruce Medaris, Major General, USA. Retired*. Plainfield, N.J.: Logos International, n.d.

Hartmann, Robert T. "U.S. to Launch Space Satellite; Will Share Date With Russia." *Los Angeles Times* July 30, 1955.

Harvey, Brian. "I Was Last to Touch the Sputnik." *Spaceflight*, January 1991.

Harwood, William B. *Raise Heaven and Earth: The Story of Martin Marietta People and Their Pioneering Achievements*. New York: Simon and Schuster, 1993.

Healey, Denis. *The Time of My Life*. New York: Norton, 1990.

Henry, Frank. "The Man Who A-Bombed Space." *Baltimore Sun*, June 7, 1959.

Henry, Thomas R. "Inventor Visions Flight to Moon." *Washington Evening Star*, October 5, 1932.

Heppenheimer, T. A. *Countdown: A History of Space Flight*. New York: Wiley, 1997.

———. "Lost in Space: What Went Wrong with NASA?" *American Heritage*, July 1992.

Herring, Mack R. *Way Station to Space: A History of the John C. Stennis Space Center*. Washington, D.C.: National Aeronautics and Space Administration, 1997. (SP-4310)

Heyn, Ernest V. *One Hundred Years of Popular Science: A Century of Wonders*. Garden City, N.Y.: Doubleday, 1972.

Hickam, Homer H., Jr. *Rocket Boys: A Memoir*. New York: Delacorte Press, 1998.

Higgins, William S. "Hermann Oberth: Founding Father of Space." *Ad Astra*, March 1990.

Hippel, Arthur R. von. "Answers to Sputnik?" *Bulletin of the Atomic Scientists*, March 1958.

Hitchcock, Lieutenant Colonel Walter T., ed. *The Intelligence Revolution: A Historical Perspective*. Washington, D.C.: Office of Air Force History, 1991. (Proceedings of the Thirteenth Military History Symposium, U.S. Air Force Academy, Colorado Springs, Colo., October 12–14, 1988.)

Holloway, Jean. *Edward Everett Hale: A Biography*. Austin: University of Texas Press, 1956.

Holquist, Michael. "The Philosophical Bases of Soviet Space Exploration." *Key Reporter,* winter 1985–86.

Holton, Sean. "Sending Dreams, Fears into Orbit: The Surrender of Germany' s von Braun Brothers Set the Stage for the Race into Space." *Orlando Sentinel,* April 30, 1995.

Hoopes, Townsend. *The Devil and John Foster Dulles.* Boston: Atlantic Monthly Press, 1973.

Hopwood, Fred. "A Week in 1958 Was a Bummer for Launches." *Florida Today* (Melbourne), September 1, 1998.

Hoyle, Fred. *Frontiers of Astronomy.* New York: Harper and Brothers, 1955.

Hunt, Linda. *Secret Agenda: The United States Government, Nazi Scientists, and Project Paperclip, 1945–1990.* New York: St. Martin's Press, 1991.

Jaffe, Leonard. *Communications in Space.* New York: Holt, Rinehart and Winston, 1966.

Jastrow, Robert. "The New Soviet Arms Buildup in Space." *New York Times Magazine,* October 3, 1982.

———. *Red Giants and White Dwarfs.* New York: Norton, 1979.

———. *Until the Sun Dies.* New York: Norton, 1977.

Jeffreys-Jones, Rhodri. *The CIA and American Democracy.* New Haven, Conn.: Yale University Press, 1989.

Johnson, Gaylord, and Irving Adler. *Discover the Stars: A Beginner's Guide to Astronomy and the Earth Satellite Project.* Revised ed. New York: Sentinel Books, 1957.

Johnson, Lyndon Baines. *The Vantage Point: Perspectives on the Presidency, 1963–1969.* New York: Holt, Rinehart and Winston, 1971.

Johnson, Nicholas L. *The Soviet Reach for the Moon.* Washington, D.C.: Cosmos Books, 1994.

Jonas, Carl. *The Sputnik Rapist.* New York: Norton, 1973.

Jones, Welton. "Vandemonium." *San Diego Union-Tribune,* July 3, 1994.

Kaplan, Joseph, Wernher von Braun, et al. *Across the Space Frontier.* New York: Viking Press, 1952. (Book form of the series of *Collier's* magazine articles that first confronted the U.S. public with the imminent possibility of manned spaceflight.)

Kauffman, James L. *Selling Outer Space: Kennedy, the Media, and Funding for Project Apollo, 1961–1963.* Tuscaloosa: University of Alabama Press, 1994.

Kaufmann, William J., III. *Discovering the Universe.* New York: Freeman, 1987.

Kearns, Doris. *Lyndon Johnson and the American Dream.* New York: Harper and Row, 1976.

Kennedy, Paul. *The Rise and Fall of the Great Powers.* New York: Random House, 1987.

Kernan, Michael. "The Man Who Launched the Space Age." *Washington Post,* September 19, 1982.

Khrushchev, Nikita. *Khrushchev Remembers: The Last Testament.* Boston: Little, Brown, 1974.

Killian, James Rhyne. *Sputnik, Scientists, and Eisenhower: A Memoir of the First Special Assistant to the President for Science and Technology.* Cambridge: MIT Press, 1977.

King, Stephen. *Danse Macabre.* New York: Everest House, 1981.

Kluger, Jeffrey. "Robert Goddard." *Time,* March 29, 1999.

Kluger, Richard. *The Paper: The Life and Death of the New York Herald-Tribune.* New York: Vintage Books, 1989.

Krieger, F. J. *Behind the Sputniks: Survey of Soviet Space Science.* Washington, D.C.: Public Affairs Press, 1958.

Krug, Linda T. *Presidential Perspectives on Space Exploration: Guiding Metaphors from Eisenhower to Bush.* New York: Prager, 1991.

Kunkel, Horst von. "The ex-Nazi Who Runs Our Space Program." *Confidential,* October 1960.

LaFeber, Walter. *America, Russia, and The Cold War, 1945–1966.* New York: Wiley, 1967.

LaFee, Scott. "Forty-one Years Ago, *Collier's* Changed the Popular View of Space." *San Diego Union-Tribune,* May 26, 1993.

Lambright, W. Henry. *Governing Science and Technology.* New York: Oxford University Press, 1976.

———. *Powering Apollo: James E Webb of NASA.* Baltimore: Johns Hopkins University Press, 1995.

Lang, Daniel. "A Reporter at Large: What's Up There?" *New Yorker,* July 31, 1948.

Lapidus, Robert J. "Sputnik and Its Repercussions: A Historical Catalyst." *Aerospace Historian,* summer–fall 1970.

Lapp, Ralph E. *Man and Space.* London: Scientific Book Club, 1961.

Larson, Arthur. *Eisenhower: The President Nobody Knew.* New York: Scribner's, 1968.

Lasby, C. *Operation Paperclip.* New York: Atheneum, 1971.

Lashmar, Paul. *Spy Flights of the Cold War.* Annapolis, Md.: Naval Institute Press, 1996.

Lasser, David A. "By Rocket to the Planets." *Nature,* November 1931.

Launius, Roger D. *Apollo: A Retrospective Analysis.* Monographs in Aerospace History, No. 3. Washington, D.C.: National Aeronautics and Space Administration, 1994.

———. "Eisenhower, Sputnik, and the Creation of NASA: Technological Elites and the Public Policy Agenda." *Prologue: Quarterly of the National Archives and Records Administration,* summer 1996.

———. *Frontiers of Space Exploration.* "Critical Events in the Twentieth Century" series. Westport, Conn.: Greenwood Press, 1998.

———. *NASA: A History of the U.S. Civil Space Program.* Malabar, Fla.: Krieger, 1994.

———. "Sputnik, the Dawn of the Space Age." *Space Times,* November–December 1997.

———, ed. *History of Rocketry and Astronautics.* AAS History Series, vol. II. San Diego: Univelt, 1994.

———, ed. *Organizing for the Use of Space: Historical Perspectives on a Persistent Issue.* AAS History Series, vol. 18. San Diego: Univelt, 1995.

Launius, Roger D., and J. D. Hunley, compilers. *An Annotated Bibliography of the Apollo Program.* Monographs in Aerospace History, No. 2. Washington, D.C.: National Aeronautics and Space Administration, 1994.

Launius, Roger D., John M. Logsdon, and Robert W. Smith, eds. *Reconsidering Sputnik: Forty Years Since the Soviet Satellite.* Amsterdam, the Netherlands: Harwood Academic Publishers, Overseas Publishers Association, 2000.

Launius, Roger D., and Howard E. McCurdy, eds. *Presidential Leadership in the Development of the U.S. Space Program.* Washington, D.C.: National Aeronautics and Space Administration, 1994. (HHR-59)

Laurence, William L. "Two Rocket Experts Argue 'Moon' Plan." *New York Times,* October 14, 1952.

Lear, John. "The Moon That Refused to Be Eclipsed." *Saturday Review,* March 5, 1960.

Lebedev, B., B. Lyk'yanov, and A. Romanov. *Sons of the Blue Planet.* New Delhi: Amerind Publishing, 1973.

Lehman, Milton. "So We Issued a Stamp—Robert Goddard Invented Rocketry Fifty Years Ago but Only Germans and Russians Paid Heed." *Washington Post,* August 22, 1955.

———. *This High Man.* New York: Farrar, Straus, 1963.

Leonard, Jonathan. *Flight into Space.* New York: Modern Library (Random House), 1953 (revised 1957).
LePage, Andrew J. "Project Orbiter: Prelude to America's First Satellite." *Spaceviews,* January 1998.
Lethbridge, Cliff. *The Story of Explorer 1: America's First Satellite.* Cape Canaveral Air Station, Fla.: Air Force Space and Missile Foundation, 1998.
Levine, Alan J. *The Missile and the Space Race.* Westport, Conn. and London: Praeger, 1994.
Levy, Lillian. *Space: Its Impact on Man and Society.* New York: Norton, 1965.
Lewis, Richard S. *Appointment on the Moon.* New York: Viking Press, 1968.
Ley, Willy. *Events in Space.* New York: David McKay, 1969.
———. *Is U.S. Building a New Moon? Popular Science,* May 1949.
———. *Missiles, Moonprobes, and Megaparsecs.* New York: Signal Science Library Books, 1964.
———. *Rockets, Missiles, and Men in Space.* New York: Viking Press, 1968.
———. *Rockets, Missiles, and Space Travel (with Sputnik Data).* New York: Viking Press, 1958.
Logsdon, John M. *The Decision to Go to the Moon: Project Apollo and the National Interest.* Chicago: University of Chicago Press, 1970.
———, ed., with Linda J. Lear, Jannelle Warren Findley, Ray A. Williamson, and Dwayne A. Day. *Exploring the Unknown: Selected Documents in the History of the U.S. Civil Space Program,* Vols. 1–4. Washington, D.C.: National Aeronautics and Space Administration, 1995–99. (NASA SP-4407)
Lonnquest, John C., and David F. Winker. *To Defend and Deter: The Legacy of the United States Cold War Missile Program.* Washington, D.C.: Department of Defense, Legacy Resource Management Program, Cold War Project, 1996.
Luongo, Kenneth N., and W. Thomas Wander. *The Search for Security in Space.* Ithaca, N.Y.: Cornell University Press, 1989.
Lovell, Bernard. *Astronomer by Chance.* New York: Basic Books, 1990.
———. *The Story of Jodrell Bank.* New York: Harper and Row, 1968.
Lovell, Jim, and Jeffrey Kluger. *Lost Moon: The Perilous Voyage of Apollo 13.* Boston: Houghton Mifflin, 1994.
Lucas, Jim G. "German Rocket Expert at Redstone Going Home—There, Security!" Scripps-Howard wire dispatch, July 3, 1957.
Lule, Jack. "Roots of the Space Race: Sputnik and the Language of U.S. News in 1957." *Journalism Quarterly* 68 (1991): 76–86.
Lyon, Peter. *Eisenhower: Portrait of the Hero.* Boston: Little, Brown, 1974.
Macdonald, Alastair. "Sputnik's Forlorn Birthplace Stirred by New Hopes." Reuters wire on *Yahoo News,* February 5, 1998.
Macko, Stanley. *Satellite Tracking.* New York: John F. Rider, 1962.
Maddy, Basil. *The Sputnik Furore.* Oakland, Calif.: published by the author, 1957.
Maguire, Jack. "Rebel Rockets Fizzled." *Arlington, Texas Daily News,* February 22, 1976.
Mailer, Norman. *A Fire on the Moon,* New York: New American Library, 1970.
Mallan, Lloyd. "The Big Red Lie." *True: The Man's Magazine,* May 1959.
———. *Men, Rockets, and Space Rats.* New York: Julian Messner, 1955.
Mangold, Tom. *Cold Warrior, James Jesus Angleton: The CIA's Master Spy Hunter.* New York: Simon and Schuster, 1991.
Marder, Murrey. "The Chinese Scientist We Expelled, and His Pupil in the Pentagon." *Washington Post,* June 14, 1981.

Marshack, Alexander. *The World in Space.* New York: Dell, 1958.
Martin, Cheryl L. "Space Race Propaganda: U.S. Coverage of the Soviet Sputnik in 1957." *Journalism Quarterly,* #64 1987.
Martin, Donald H. *Communication Satellites, 1958–1992.* El Segundo, Calif.: Aerospace, 1991.
Matson, Wayne R., ed. *Cosmonautics: A Colorful History of the Soviet-Russian Space Programs.* Washington, D.C.: Cosmos Books, 1994.
Matthews, J. B. "Relief for America's Reds." *American Legion Magazine,* October 1957.
McAleer, Neil. "The Space Age Turns Thirty." *Space World,* October 1987.
McCurdy, Howard E. *Inside NASA: High Technology and Organizational Change in the U.S. Space Program.* Baltimore: Johns Hopkins University Press, 1993.
———. *Space and the American Imagination.* Smithsonian History of Aviation Book Series. Washington, D.C.: Smithsonian Institution Press, 1997.
McDonough, Thomas R. *Space: The Next Twenty-five Years.* New York: Wiley, 1987.
McDougall, Walter A. *The Heavens and the Earth: A Political History of the Space Age.* New York: Basic Books, 1985.
———. "Sputnik, the Space Race, and the Cold War." *Bulletin of the Atomic Scientists* 41 (may 1985): 20–25.
McGarvey, Patrick J. *C.I.A.: The Myth & the Madness.* New York: Saturday Review Press, 1972.
Medaris, John G., and Arthur Gordon. *Countdown for Decision.* New York: Putnam's, 1960.
Menzel, Donald H., and Lyle G. Boyd. *The World of Flying Saucers: A Scientific Examination of a Major Myth of the Space Age.* Garden City, N.Y.: Doubleday, 1963.
Miller, Merle. *Lyndon.* New York: Putnam's, 1980.
Minton, Arthur. "Sputnik and Some of Its Offshootniks." *Names,* June 1958.
Moore, Patrick, ed. *The International Encyclopedia of Astronomy.* New York: Orion Books, 1987.
Morin, Relman. *Dwight D. Eisenhower: A Gauge of Greatness.* Washington, D.C.: Associated Press Books, 1969.
Morrow, Edward R. *In Search of Light: The Broadcasts of Edward R. Murrow, 1938–1961.* New York: Knopf, 1967.
Muir, Charles S. *A Trip to Polaris.* Washington, D.C.: Polaris Company, 1923.
Murray, Charles, and Catherine Bly Cox. *Apollo—the Race to the Moon.* New York: Simon and Schuster, 1989.
Narvaez, Alfonso A. "Dr. John P. Hagan, eighty-two, Director of First Major U.S. Space Effort" (obituary). *New York Times,* September 1, 1990.
Nash, Philip. *The Other Missiles of October: Eisenhower, Kennedy, and the Jupiters, 1957–1963.* Chapel Hill: University of North Carolina Press, 1997.
National Aeronautics and Space Administration History Office. *Documents in the History of NASA: An Anthology.* Washington, D.C.: National Aeronautics and Space Administration, 1975.
Needell, Allan A., ed. *The First Twenty-Five Years in Space: A Symposium.* Washington, D.C.: Smithsonian Institution Press, 1983.
Neufeld, Michael J. *The Rocket and the Reich: Peenemünde and the Coming of the Ballistic Missile Era.* New York: Free Press, 1995.
Newell, Homer E. *Beyond the Atmosphere: Early Years of Space Science.* Washington, D.C.: National Aeronautics and Space Administration, 1980. (NASA SP-4211)
———. *Express to the Stars.* New York: McGraw-Hill, 1961.
———, ed. *Sounding Rockets.* New York: McGraw-Hill, 1959.

Newhouse, John. *Cold Dawn: The Story of SALT.* New York: Holt, Rinehart and Winston, 1973.
Newlon, Clarke. *The Aerospace Age Dictionary.* New York: Franklin Watts, 1975.
Newton, David E. *U.S. and Soviet Space Programs: A Comparison.* New York: Franklin Watts, 1988.
Nixon, Richard M. *Memoirs.* New York: Grosset and Dunlap, 1975.
Norton, Howard. *Only in Russia.* Princeton, N.J.: Van Nostrand, 1961.
Oakley, J. Ronald. *God's Country: America in the Fifties.* New York: Debner Books, 1986.
Oberg, James E. "Disaster at the Cosmodrome." *Air & Space/Smithsonian,* September–October 1990.
———. *Red Star in Orbit.* New York: Random House, 1981.
Oberg, James E., and Alcestis R. Obert. *Pioneering Space.* New York: McGraw-Hill, 1987.
Obst, David. *Too Good to Be Forgotten.* New York: Wiley, 1998.
O'Connor, Ulick. *Sputnik and Other Poems.* New York: Devin-Adair, 1967.
O'Keefe, Bernard J. *Nuclear Hostages.* Boston: Houghton Mifflin, 1983.
Olsen, Jack. *Aphrodite: Desperate Mission.* New York: Putnam's, 1970.
Ordway, Frederic I., III. "Will Russian Scientists Beat Us to the Moon?" *American Mercury,* February 1958.
———, and Randy Liebermann, eds. *Blueprint for Space: From Science Fiction to Science Fact.* Washington, D.C.: Smithsonian Institution Press, 1992.
Ordway, Frank I., and Mitchell R. Sharpe. *The Rocket Team.* New York: Crowell, 1979.
Osman, Tony. *Space History.* New York: St. Martin's Press: 1983.
O'Toole, Thomas. "When Sputnik Shocked Us." *Washington Post,* October 4, 1982.
Pach, Chester J., Jr., and Elmo Richardson. *The Presidency of Dwight D. Eisenhower.* Lawrence: University Press of Kansas, 1991.
Park, Robert L. *Voodoo Science: The Road from Foolishness to Fraud.* New York: Oxford University Press, 2000.
Paul, Günter. *The Satellite Spin-Off.* Washington, D.C.: Robert B. Luce, 1975.
Pedlow, Gregory W., and Donald W. Welzenbach. *The CIA and the U-2 Program, 1954–1974.* Washington, D.C.: Central Intelligence Agency, 1998.
Peebles, Curtis. *The Battle for Space.* New York: Beaufort Books, 1983.
———. *The Corona Project: America's First Spy Satellites.* Annapolis, Md.: United States Naval Institute, 1997.
———. *Guardians: Strategic Reconnaissance Satellites.* Novato, Calif.: Presidio, 1987.
———. *The Moby Dick Project, Reconnaissance Balloons over Russia.* Washington, D.C.: Smithsonian Institution Press, 1991.
———. *Watch the Skies! A Chronicle of the Flying Saucer Myth.* Washington, D.C.: Smithsonian Institution Press, 1994.
Pellegrino, Charles R., and Joshua Stoff. *Chariots for Apollo.* New York: Atheneum, 1985.
Pendray, G. Edward. "Dr. Robert H. Goddard: The Man Who Ushered in the Space Age." *American Legion,* September 1960.
———. "He Shot a Rocket and Reaped Scorn." *Washington Post,* March 11, 1962.
Perret, Geoffrey. *Eisenhower.* New York: Random House, 1999.
Persico, Joseph E. *Casey: From the OSS to the CIA.* New York: Viking Penguin, 1990.
Philips, Alan. "Down-to-Earth Pupils Spurn Final Frontier: Children Prefer a Career as a Gangster to Being a Cosmonaut." *Daily Telegraph,* October 6, 1997.
Pincus, Walter. "Military Got Authority to Use Nuclear Arms in 1957." *Washington Post,* March 21, 1998.

Pincus, Walter, and George Lardner Jr. "Eisenhower Issued Limited Nuclear Authority Absent Presidential Order." *Washington Post,* September 2, 1998.
Piszkiewicz, Dennis. *Wernher von Braun: The Man Who Sold the Moon.* Westport, Conn.: Praeger, 1998.
Powell, Joel W. "Juno I: America's First Space Launch Vehicle." *Quest* 4, no. 4 (winter 1995): 12–13.
Portree, David S. F. *Mir Hardware Heritage.* Washington, D.C.: National Aeronautics and Space Administration, March 1995. (NASA Reference Publication 1357)
———. *NASA's Origins and the Dawn of the Space Age.* NASA Monographs in Aerospace History, No. 10. Washington, D.C.: National Aeronautics and Space Administration, 1998.
Prados, John. *The Soviet Estimate.* New York: Dial Press, 1982.
Radosh, Ronald, and Joyce Milton. *The Rosenberg File: A Search for the Truth.* New York: Holt, Rinehart and Winston, 1983.
Ramo, Simon. *The Business of Science: Winning and Losing in the High-Tech Age.* New York: Hill and Wang, 1988.
RAND Corporation. *The RAND Corporation: The First Fifteen Years.* Santa Monica, Calif., 1963.
Randles, Jenny, and Peter Hough. *The Complete Book of UFO's.* New York: Sterling, 1996.
Ranelagh, John. *The Agency: The Rise and Fall of the CIA.* New York: Simon and Schuster, 1986.
Rasmussen, Fred. "Remember When." *Baltimore Sun,* October 12, 1997.
Ravitch, Diane. *Left Back: A Century of Failed School Reforms.* New York: Simon and Schuster, 2000.
Reed, Fred. "The Day the Rocket Died." *Air & Space,* October 1987.
Reiffel, Leonard. "Sagan Breached Security by Revealing U.S. Work on a Lunar Bomb Project." *Nature,* May 4, 2000.
Rhea, John, ed. *Roads to Space.* New York: Aviation Week Group, 1995.
Riabchikov, Evgeny. *Russians in Space.* Garden City, N.Y.: Doubleday, 1971.
Rich, Ben R., and Leo Janos. *Skunk Works.* Boston: Little Brown, 1994.
Richelson, Jeffrey T. *America's Secret Eyes in Space: The U.S. Keyhole Spy Satellite.* New York: HarperBusiness, 1990.
Ridpath, Ian. *A Dictionary of Astronomy.* New York: Oxford University Press, 1998.
Roberts, Chalmers. *First Rough Draft.* New York: Praeger, 1973.
———. *The Nuclear Years: The Arms Race and Arms Control, 1945–1970.* New York: Praeger, 1970.
Robinson, Jeffrey. *The End of the American Century: Hidden Agendas of the Cold War.* London: Hutchinson, 1992.
Roland, Alex. *A Space-Faring People: Perspectives on Early Spaceflight.* Washington, D.C.: National Aeronautics and Space Administration, 1985. (NASA SP-4405)
Roman, Peter J. *Eisenhower and the Missile Gap.* Ithaca, N.Y.: Cornell University Press, 1995.
Romanov, A. *Spacecraft Designer: The Story of Sergey Korolev.* Moscow: Novosti Press Agency, 1976.
Rosen, M. W. "October 4, 1957: Sputnik Beeps, America Responds." *Astronautics and Aeronautics* 15 (October 1977): 20–23.
———. *The Viking Rocket Story.* New York: Harper and Brothers, 1955.
Rosholt, Robert L. *An Administrative History of NASA, 1955–1965.* Washington, D.C.:

National Aeronautics and Space Administration, 1966. (NASA SP-4101)

Rostow, W. W. *The Diffusion of Power: An Essay in Recent History.* New York: Macmillan, 1972.

———. *Open Skies: Eisenhower's Proposal of July 21, 1955.* Austin: University of Texas Press, 1982.

Ruffner, Kevin. *Corona: America's First Satellite Program.* Washington, D.C.: Center for the Study of Intelligence, 1995.

"Satellite Progress." *Time,* October 15, 1956.

Schecter, Jerrold, and Peter Deriabin. *The Spy Who Saved the World: How a Soviet Colonel Changed the Course of the Cold War.* New York: Scribner's, 1992.

Scheer, Julian. "Sputnik Launched Space Age Forty Years Ago: Russian Triumph Comes Full Circle." *USA Today,* October 2, 1997.

Schefter, James. *The Race: The Uncensored Story of How America Beat Russia to the Moon.* Garden City, N.Y.: Doubleday, 1999.

Schneir, Walter, and Miriam Schneir. *Invitation to an Inquest.* Garden City, N.Y.: Doubleday, 1965.

Scully, Frank. *Behind the Flying Saucers.* New York: Henry Holt, 1950.

Seamans, Robert C., Jr. *Aiming at Targets: The Autobiography of Robert C. Seamans, Jr.* Washington, D.C.: National Aeronautics and Space Administration, 1996. (NASA SP-4106)

Sears, Donald A., and Henry A. Smith. "A Linguistic Look at Aerospace English." *Air Force / Space Digest,* December 1969.

"The Seer of Space." *Life,* November 18, 1957.

Shaw, Captain John. "The Influence of Space Power upon History (1944–1998)." *Air Chronicles,* March 1999.

Shelton, William. *Soviet Space Exploration.* New York: Washington Square Press, 1968.

Shelton, William Roy. *Countdown: The Story of Cape Canaveral.* Boston: Little Brown, 1960.

Shepard, Alan, and Deke Slayton. *Moon Shot: The Inside Story of America's Race to the Moon.* Atlanta: Turner Publishing, 1994.

Sherry, Michael S. *In the Shadow of War: The United States Since the 1930s.* New Haven, Conn.: Yale University Press, 1995.

Shirer, William. *The Rise and Fall of the Third Reich.* New York: Simon and Schuster, 1960.

Shternfeld, Ari. *Soviet Space Science.* New York: Basic Books, 1959.

Shuriden, G. A. *Mastery of Outer Space in the USSR, 1957–1967.* Washington, D.C.: National Aeronautics and Space Administration, 1975.

Siddiqi, Asif A. "Before Sputnik: Early Satellite Studies in the Soviet Union, 1947–1957." *Spaceflight,* October and November 1997.

———. "Soviet Space Program." *Spaceflight,* August 1994.

Sistak, B. J. "Memories of Sputnik." *Astronomy* 14 (December 1986).

Slotkin, Richard. *Gunfighter Nation: The Myth of the Frontier in Twentieth-Century America.* New York: Atheneum, 1992.

Smal-Stocki, Roman. *The Impact of the "Sputnik" on the English Language of the U.S.A.* Chicago: Shevchenko Scientific Society Study Center, 1958.

Smith, Bruce L. R. *The Rand Corporation.* Cambridge: Harvard University Press, 1966.

Smolders, Peter. *Soviets in Space.* New York: Taplinger, 1971.

Snow, C. P. *Science and Government.* Cambridge: Harvard University Press, 1960.

———. *The Two Cultures and the Scientific Revolution.* New York: Cambridge University Press, 1959.
Special Issue: Twenty-Five Years of Space Exploration. Sky and Telescope, October 1982.
Spires, David N. *Beyond Horizons: A Half Century of Air Force Space Leadership.* Peterson Air Force Base, Colo: Air Force Space Command, 1997.
Srodes, James. *Allen Dulles: Master of Spies.* Washington, D.C.: Regnery, 1999.
Stares, Paul B. *The Militarization of Space: U.S. Policy, 1945–1984.* Ithaca, N.Y.: Cornell University Press, 1985.
Stehling, Kurt K. *Project Vanguard.* Garden City, N.Y.: Doubleday, 1961.
Stimson, Thomas E. "He Spies on Satellites." *Popular Mechanics,* October 1955.
Stine, Harry. *Earth Satellites and the Race for Space Superiority.* New York: Ace Books, 1957.
Stoiko, Michael. *Soviet Rocketry: Past, Present and Future.* New York: Holt, Rinehart and Winston, 1970.
Stuhlinger, Ernst, and Frederic I. Ordway, III. *Wernher von Braun: Crusader for Space.* 2 vols. Malabar, Fla.: Krieger, 1994.
Suid, Lawrence. "Kennedy, Apollo, and the Columbus Factor." *Spaceflight,* July 1994.
Sullivan, Walter. *Assault on the Unknown: The International Geophysical Year.* New York: McGraw-Hill, 1961.
Swenson, Lloyd S., Jr., James M. Grimwood, and Charles C. Alexander. *This New Ocean: A History of Project Mercury.* Washington, D.C.: National Aeronautics and Space Administration, 1966. (NASA SP-4201)
Szulc, Tad. *Then and Now: How the World Has Changed Since World War II.* New York: William Morrow, 1990.
Talbott, Strobe. ed., *Khrushchev Remembers.* Toronto, Canada, and Boston: Little, Brown, 1970.
Taylor, Henry J. "Dr. Goddard's Rockets Pointed Way to Space." *Birmingham Post-Herald,* August 17, 1973.
Taylor, Maxwell D. *The Uncertain Trumpet.* New York: Harper and Brothers, 1960.
Thomas, Shirley. *Men of Space.* Vols. 1–7. Philadelphia: Chilton, 1960–65.
Tikhonravov, M. K. "The Creation of the First Artificial Earth Satellite: Some Historical Details." *Journal of the British Interplanetary Society* 47, no. 5 (May 1994): 191–94.
Toffler, Alvin, ed. *The Futurists.* New York: Random House, 1972.
Trench, Brinsley Le Poer. *The Flying Saucer Story.* London: Neville Spearman, 1966.
Trewhitt, Henry L. *McNamara.* New York: Harper and Row, 1971.
Tuohy, John. "Excitement of First Satellite Orbit Relived." *Florida Today* (Melbourne), February 1, 1998.
U.S. Congress. Senate. Committee on Armed Services. *Inquiry into Satellite and Missile Programs: Hearings Before the Preparedness Investigating Subcommittee of the Committee on Armed Services.* 85th Congr., three sessions, parts 1, 2, and 3.
U.S. Congress. Senate. Committee on Commerce, Science, and Transportation. *Soviet Space Programs, 1976–80, Unmanned Space Activities.* May 1985.
Vakhnin, V. "Artificial Earth Satellites." *QST,* November 1957.
Van Allen, James A., ed. *Scientific Uses of Earth Satellites.* Ann Arbor: University of Michigan Press, 1956.
Vedeshin, L. S., and V. P. Dudykin. "Preparation and Launching in the USSR of the First Artificial Earth Satellite." *ITU Telecommunication Journal* 44 (October 1977): 477–481. (Microfiche 78A17750)
Wainwright, Loudon. "How Insiders Kept Their Great Secret." *Life,* January 5, 1959.

Wallace, Lane E. *Dreams, Hopes, Realities: NASA's Goddard Space Flight Center—The First Forty Years.* Washington, D.C.: National Aeronautics and Space Administration, 1999. (NASA SP-4312)

Walter, Chip. "Tsolkovsky's Gift." *Final Frontier.* July–August 1991.

Walter, William J., *Space Age.* New York: Random House, 1992.

Walters, Helen B. *Wernher von Braun: Rocket Engineer.* New York: Macmillan, 1964.

Walters, Vernon A. *Silent Missions.* Garden City, N.Y.: Doubleday, 1978.

Ward, Bob. *The Light Stuff—Space Humor—From Sputnik to the Shuttle.* Huntsville, Ala.: Jester Books, 1982.

Wasserman, Harvey, and Norman Solomon, with Robed Alvarez and Eleanor Walters. *Killing Our Own: Chronicling the Disaster of America's Experience with Atomic Radiation, 1945–1982.* New York: Delta Books, 1982.

Webb, James E. *Space Age Management.* New York: McGraw-Hill, 1962.

Wells, Helen T., Susan B. Whiteley, and Carrie Karegeannes. *Origins of NASA Names.* Washington, D.C.: National Aeronautics and Space Administration, 1976. (NASA SP-4402)

Werth, Alexander. *Russia Under Khrushchev.* New York: Crest Books, 1962.

Westwood, J. N. *Russia, 1917–1964.* New York: Harper and Row, 1966.

White, Frank. *The Overview Effect.* Boston: Houghton Mifflin, 1987.

Wilford, John Noble. *We Reach the Moon.* New York: Bantam, 1969.

Williams, Beryl, and Samuel Epstein. *The Rocket Pioneers.* New York: Julian Messner, 1977.

Williams, Gurney. "Sputnik: The Little Sphere That Changed the World." *Popular Mechanics* 164 (October 1987): 59–61.

Wilson, Eugene Edward. *Kitty Hawk to Sputnik to Polaris; a Contemporary Account of the Struggle over Military and Commercial Air Policy in the United States.* Palm Beach, Fla.: Literary Investors Guild, 1967.

Wilson, Glen P. "The Legislative Origins of NASA: The Role of Lyndon B. Johnson." *Prologue: Quarterly of the National Archives,* winter 1993.

Winks, Robin W. *The Historian as Detective.* New York: Harper Colophon, 1970.

Winslow, Richard K. "Race into Space: Can We Win?" *Newsweek,* March 4, 1957.

Winter, Frank H. *The First Golden Age of Rocketry: Congreve and Hale Rockets of the Nineteenth Century.* Washington, D.C.: Smithsonian Institution Press, 1990.

———. *Rockets into Space.* Cambridge: Harvard University Press, 1990.

Withers, A. M. "Words and the Space Age." *Word Study* (Merriam-Webster). February 1962.

Witkin, Richard, ed. *The Challenge of the Sputniks.* Garden City, N.Y.: Doubleday, 1958.

Wolfe, Tom. *The Right Stuff.* New York: Farrar, Straus and Giroux, 1979.

Woodbury, David O. *Around the World in Ninety Minutes.* New York: Harcourt Brace, 1958.

Wright, Mike. "The Disney–Von Braun Collaboration and Its Influence on Space Exploration." In *Inner Space, Outer Space: Humanities, Technology, and the Postmodern World,* ed. Daniel Schenker, Craig Hanks, and Susan Kray. Huntsville, Ala.: Southern Humanities Conference, 1993.

———. "Huntsville and the Space Program." *Alabama Heritage,* spring and summer 1998. (Two-part article)

Yaroslav, Golovanov, and Sergei Korolev. *The Apprenticeship of a Space Pioneer.* Moscow: Mir Publishers, 1975.

Yenne, Bill. *Pictorial History of World Spacecraft*. New York: Exeter Books, 1988.
Yoder, Edwin M., Jr. *Joe Alsop's Cold War*. Chapel Hill: University of North Carolina Press, 1995.
Zheleznov, Nikolai. "Hello . . . Bip-Bip." *Soviet Life*, October 1982.
Zubok, Vladislav, and Constatine Pleshakov. *Inside the Kremlin's Cold War: From Stalin to Khrushchev*. Cambridge: Harvard University Press, 1996.
Zwicky, Fritz. *Report on Certain Phases of War Research in Germany*. Dayton, Ohio: Air Material Command, 1947.

OTHER MEDIA

Unpublished interviews. A number of interviews were invaluable and are on file at the NASA History Office in Washington, D.C., including those with Andrew J. Goodpaster Jr. and Bryce N. Harlow.

Video transcript. The transcript of an interview with Major General John B. Medaris on *Meet the Press*, February 9, 1953, provided valuable information. It was produced by Lawrence E. Spivak.

Internet. The single most useful tool used in preparing this book was Google Uncle Sam (Google.com—go to Uncle Sam), which allows one to sift through an infinite number of federal government sites at once. As new information appears on space history, the most valuable sites are the "Space Online" at *Florida Today* (http://www.flatoday.com), the NASA History Office (http://www.hq.nasa.gov/office/pao/History), Spaceviews (http://www.seds.org/spaceviews), and Space.com (http://www.space.com).

General information on space is best accessed through http://www.nasa.gov

The negative side of Internet access is that it has been used to totally replace many printed references, giving them a new ephemeral nature. "Sorry, we don't have any paper copies of the information you will find on our Web site," writes Mark C. Cleary, historian at Patrick Air Force Base, Florida. "The books went out of publication in 1994, and the commander doesn't have money in the budget to print any more." The site in question (pafb.af.mil/heritage/heritage.htm#45) is very difficult to find even with the most sophisticated search engine, and it stands to reason that the next commander may decide to remove the information from the larger site for lack of bandwidth or interest.

Conferences. During the course of researching this book, I was fortunate to be able to attend two conferences which were of great value. The first was "Space Exploration at the Millennium," held on March 24, 1999, at American University in Washington, D.C. The second, "Developing U.S. Launch Capability: The Role of Civil-Military Cooperation," took place on November 5, 1999, at the American Association for the Advancement of Science in Washington and allowed one to hear such pioneers as Dr. Simon Ramo and General Bernard Schriever. Unfortunately, neither of these important events was followed by a transcript or even a series of abstracts, so they have been rendered ephemeral for their lack of documentation.

Exhibit. "Space Race," which opened on May 16, 1997, at the National Air and Space Museum in Washington, D.C., gave me the chance to meet and discuss Sputnik with Ross Perot, Konstantin Feokistov, Aleksei Leonov, and German Titov. It also offered a rare opportunity to see the artifacts of the early Space Age, when success was measured by headline-making "firsts": Sputnik's arming key, the Vanguard TV3 satellite, a Jupiter C nose cone, and a Corona KH-4B camera and film return capsule.

Index